Methods in Cell Biology

Microfluidics in Cell Biology
Part B: Microfluidics in Single Cells

Volume 147

Series Editors

Leslie Wilson
Department of Molecular, Cellular and Developmental Biology
University of California
Santa Barbara, California

Phong Tran
University of Pennsylvania
Philadelphia, USA &
Institut Curie, Paris, France

Methods in Cell Biology

Microfluidics in Cell Biology
Part B: Microfluidics in Single Cells

Volume 147

Edited by

Matthieu Piel

Systems Biology of Cell Division and Cell Polarity, Cell Biology and Cancer Department, Institut Curie; Institut Pierre Gilles de Gennes for Microfluidics, Paris, France

Daniel Fletcher

Bioengineering & Biophysics, University of California, Berkeley; Biological Systems & Engineering, Lawrence Berkeley National Laboratory, Berkeley; Chan Zuckerberg Biohub, San Francisco, CA, United States

Junsang Doh

Department of Mechanical Engineering/School of Interdisciplinary Bioscience and Bioengineering (I-Bio), Pohang University of Science and Technology (POSTECH), Pohang, South Korea

ACADEMIC PRESS

An imprint of Elsevier

Academic Press is an imprint of Elsevier
50 Hampshire Street, 5th Floor, Cambridge, MA 02139, United States
525 B Street, Suite 1650, San Diego, CA 92101, United States
The Boulevard, Langford Lane, Kidlington, Oxford OX5 1GB, United Kingdom
125 London Wall, London, EC2Y 5AS, United Kingdom

First edition 2018

Notices
Knowledge and best practice in this field are constantly changing. As new research and
experience broaden our understanding, changes in research methods, professional practices, or
medical treatment may become necessary.

Practitioners and researchers must always rely on their own experience and knowledge in
evaluating and using any information, methods, compounds, or experiments described herein.
In using such information or methods they should be mindful of their own safety and the safety
of others, including parties for whom they have a professional responsibility.

To the fullest extent of the law, neither the Publisher nor the authors, contributors, or editors,
assume any liability for any injury and/or damage to persons or property as a matter of products
liability, negligence or otherwise, or from any use or operation of any methods, products,
instructions, or ideas contained in the material herein.

ISBN: 978-0-12-814282-0
ISSN: 0091-679X

For information on all Academic Press publications
visit our website at https://www.elsevier.com/books-and-journals

Working together
to grow libraries in
developing countries

www.elsevier.com • www.bookaid.org

Publisher: Zoe Kruze
Acquisition Editor: Zoe Kruze
Editorial Project Manager: Teresa Pons-Ferrer
Production Project Manager: Denny Mansingh
Cover Designer: Miles Hitchen

Typeset by SPi Global, India

Contents

SECTION 3 MICROFLUIDICS FOR CELL MECHANICS

Contributors

Koceila Aizel
Laboratoire Colloïdes et Matériaux Divisés, CNRS UMR 8231, Chemistry Biology & Innovation, ESPCI Paris, PSL Research University, Paris, France

Lucile Alexandre
Laboratoire Phyisico Chimie Curie, Institut Curie, PSL Research University; Sorbonne Universités, UPMC Univ Paris 06; Institut Pierre-Gilles de Gennes, Paris, France

Robert H. Austin
Department of Physics, Princeton University, Princeton, NJ, United States

Julien Babic
University of Rennes, Institute of Genetics and Development, CNRS, Rennes, France

Amel Bendali
Laboratoire Phyisico Chimie Curie, Institut Curie, PSL Research University; Sorbonne Universités, UPMC Univ Paris 06; Institut Pierre-Gilles de Gennes, Paris, France

Alka Bhat
Laboratory of Cell Physics ISIS/IGBMC, CNRS and University of Strasbourg, Strasbourg; Institut de Génétique et de Biologie Moléculaire et Cellulaire; Université de Strasbourg; Centre National de la Recherche Scientifique, UMR7104; Institut National de la Santé et de la Recherche Médicale, U964, Illkirch, France

Jérôme Bibette
Laboratoire Colloïdes et Matériaux Divisés, CNRS UMR 8231, Chemistry Biology & Innovation, ESPCI Paris, PSL Research University, Paris, France

Zuzana Bilkova
Department of Biological and Biochemical Sciences, Faculty of Chemical Technology, University of Pardubice, Pardubice, Czech Republic

Julia Bos
Pasteur Institute, Department of Genomes and Genetics, Paris, France

Nicolas Bremond
Laboratoire Colloïdes et Matériaux Divisés, CNRS UMR 8231, Chemistry Biology & Innovation, ESPCI Paris, PSL Research University, Paris, France

Gilles Charvin
Department of Developmental Biology and Stem Cells, Institut de Génétique et de Biologie Moléculaire et Cellulaire; Centre National de la Recherche Scientifique, UMR7104; Institut National de la Santé et de la Recherche Médicale; Université de Strasbourg, Illkirch, France

Andrew G. Clark
Institut Curie, PSL Research University, CNRS, UMR 144, Paris, France

Damien Coudreuse
University of Rennes, Institute of Genetics and Development, CNRS, Rennes, France

Shane Deegan
Aquila Bioscience limited, Business Innovation Centre; Glycoscience Group, National Centre for Biomedical Engineering Science, National University of Ireland Galway, Galway, Ireland

M. Delarue
MILE, Laboratory for Analysis and Architecture of Systems, CNRS, Toulouse, France

Matthieu Delincé
School of Life Sciences, Swiss Federal Institute of Technology in Lausanne (EPFL), Lausanne, Switzerland

Stéphanie Descroix
Laboratoire Phyisico Chimie Curie, Institut Curie, PSL Research University; Sorbonne Universités, UPMC Univ Paris 06; Institut Pierre-Gilles de Gennes, Paris, France

Bruno Dupuy
Laboratory Pathogenesis of Bacterial Anaerobes, Department of Microbiology, Institut Pasteur, Paris, France

Felix Ellett
BioMEMS Resource Center, Department of Surgery, Massachusetts General Hospital, Harvard Medical School, Shriners Burns Hospital, Boston, MA, United States

Youlian Goulev
Department of Developmental Biology and Stem Cells, Institut de Génétique et de Biologie Moléculaire et Cellulaire; Centre National de la Recherche Scientifique, UMR7104; Institut National de la Santé et de la Recherche Médicale; Université de Strasbourg, Illkirch, France

Jochen Guck
Biotechnology Center, Center for Molecular and Cellular Bioengineering, Technische Universität Dresden, Dresden, Germany

O. Hallatschek
Department of Physics and Integrative Biology, University of California, Berkeley, CA, United States

L.J. Holt
Institute for Systems Genetics, New York University Langone Health, New York, NY, United States

Daniel Irimia
BioMEMS Resource Center, Department of Surgery, Massachusetts General
Hospital, Harvard Medical School, Shriners Burns Hospital, Boston, MA,
United States

Lokesh Joshi
Aquila Bioscience limited, Business Innovation Centre; Glycoscience Group,
National Centre for Biomedical Engineering Science, National University of
Ireland Galway, Galway, Ireland

Martin Kräter
Biotechnology Center, Center for Molecular and Cellular Bioengineering,
Technische Universität Dresden, Dresden, Germany

Emilie Le Maout
Laboratory of Cell Physics ISIS/IGBMC, CNRS and University of Strasbourg,
Strasbourg; Institut de Génétique et de Biologie Moléculaire et Cellulaire;
Université de Strasbourg; Centre National de la Recherche Scientifique,
UMR7104; Institut National de la Santé et de la Recherche Médicale, U964,
Illkirch, France

Alex Leithner
Institute of Science and Technology Austria (IST Austria), Klosterneuburg, Austria

Chwee Teck Lim
Department of Biomedical Engineering; Mechanobiology Institute; Biomedical
Institute for Global Health Research and Technology, National University of
Singapore, Singapore, Singapore

Simon Lo Vecchio
Laboratory of Cell Physics ISIS/IGBMC, CNRS and University of Strasbourg,
Strasbourg; Institut de Génétique et de Biologie Moléculaire et Cellulaire;
Université de Strasbourg; Centre National de la Recherche Scientifique,
UMR7104; Institut National de la Santé et de la Recherche Médicale, U964,
Illkirch, France

Laurent Malaquin
Laboratoire Phyisico Chimie Curie, Institut Curie, PSL Research University;
Sorbonne Universités, UPMC Univ Paris 06; Institut Pierre-Gilles de Gennes,
Paris, France

Anika L. Marand
BioMEMS Resource Center, Department of Surgery, Massachusetts General
Hospital, Harvard Medical School, Shriners Burns Hospital, Boston, MA,
United States

Audrey Matifas
Department of Developmental Biology and Stem Cells, Institut de Génétique et de
Biologie Moléculaire et Cellulaire; Centre National de la Recherche Scientifique,
UMR7104; Institut National de la Santé et de la Recherche Médicale; Université
de Strasbourg, Illkirch, France

John D. McKinney
School of Life Sciences, Swiss Federal Institute of Technology in Lausanne
(EPFL), Lausanne, Switzerland

Jack Merrin
Institute of Science and Technology Austria (IST Austria), Klosterneuburg, Austria

Sinan Muldur
BioMEMS Resource Center, Department of Surgery, Massachusetts General
Hospital, Harvard Medical School, Shriners Burns Hospital, Boston, MA,
United States

Javier Muñoz-Garcia
University of Rennes, Institute of Genetics and Development, CNRS, Rennes,
France

Iago Pereiro
Laboratoire Phyisico Chimie Curie, Institut Curie, PSL Research University;
Sorbonne Universités, UPMC Univ Paris 06; Institut Pierre-Gilles de Gennes,
Paris, France

Jörg Renkawitz
Institute of Science and Technology Austria (IST Austria), Klosterneuburg, Austria

Anne Reversat
Institute of Science and Technology Austria (IST Austria), Klosterneuburg, Austria

Daniel Riveline
Laboratory of Cell Physics ISIS/IGBMC, CNRS and University of Strasbourg,
Strasbourg; Institut de Génétique et de Biologie Moléculaire et Cellulaire;
Université de Strasbourg; Centre National de la Recherche Scientifique,
UMR7104; Institut National de la Santé et de la Recherche Médicale, U964,
Illkirch, France

Philipp Rosendahl
Biotechnology Center, Center for Molecular and Cellular Bioengineering,
Technische Universität Dresden, Dresden, Germany

Anthony Simon
Institut Curie, PSL Research University, CNRS, UMR 144, Paris, France

Michael Sixt
Institute of Science and Technology Austria (IST Austria), Klosterneuburg, Austria

Jana Srbova
Department of Biological and Biochemical Sciences, Faculty of Chemical
Technology, University of Pardubice, Pardubice, Czech Republic

Sanae Tabnaoui
Laboratoire Phyisico Chimie Curie, Institut Curie, PSL Research University;
Sorbonne Universités, UPMC Univ Paris 06, Paris, France

Chiara Toniolo
School of Life Sciences, Swiss Federal Institute of Technology in Lausanne (EPFL), Lausanne, Switzerland

Marta Urbanska
Biotechnology Center, Center for Molecular and Cellular Bioengineering, Technische Universität Dresden, Dresden, Germany

Ramanathan Vaidyanathan
Department of Biomedical Engineering, National University of Singapore, Singapore, Singapore

Danijela Matic Vignjevic
Institut Curie, PSL Research University, CNRS, UMR 144, Paris, France

Jean-Louis Viovy
Laboratoire Phyisico Chimie Curie, Institut Curie, PSL Research University; Sorbonne Universités, UPMC Univ Paris 06; Institut Pierre-Gilles de Gennes, Paris, France

Trifanny Yeo
Department of Biomedical Engineering, National University of Singapore, Singapore, Singapore

Preface

Microfluidics, or the control of fluids at the micron scale, refers to a recent set of techniques that grew rapidly in the 1990s out of engineering and physical science laboratories and were almost immediately applied to biology, in particular cell biology. This rapid adoption by cell biologists is primarily due to the availability of biocompatible materials that can be microfabricated to enable work with live cells and biomolecules. Remarkably, culturing cells in such devices has been straight-forward, with some hard-to-culture cells like neurons appearing to grow better in microfluidic devices than regular Petri dishes. The second reason for the rapid suc-cess of microfluidics in biology is the possibility to quantitatively and temporally control the environment of live cells and multicellular assemblies or organisms, and to perform miniaturized analytic procedures thanks to simple implementation of single-cell sorting and encapsulation. These developments have led to the general concepts of controlled cell microenvironment, lab-on-a-chip, and organ-on-a-chip devices, with a very wide range of applications, including some that have become industrial products. In these volumes of *Methods in Cell Biology*, we focus on applications of microfluidics to cell biology. Part A (Volume 146) deals with model multicellular assemblies (monolayers and spheroids), organs on chip, and model organisms; Part B (Volume 147) deals with single cells, including application to microorganisms, cell migration, and cell mechanics; and finally, Part C (Volume 148) deals with applications for cell analysis including cell culture and sorting, droplet-based microfluidics, and single-cell analysis. These volumes cover a broad range of application of microfluidics to cell biology, with detailed methods for specific cases.

Junsang Doh, Daniel A. Fletcher, and Matthieu Piel

Microfluidics for micro-organisms

Drug delivery and temperature control in microfluidic chips during live-cell imaging experiments

1

Javier Muñoz-Garcia, Julien Babic, Damien Coudreuse[1]

University of Rennes, Institute of Genetics and Development, CNRS, Rennes, France
[1]Corresponding author: e-mail address: damien.coudreuse@univ-rennes1.fr

CHAPTER OUTLINE

Methods in Cell Biology, Volume 147, ISSN 0091-679X, https://doi.org/10.1016/bs.mcb.2018.06.004

Abstract

Microfluidic technologies have become a standard tool in cell biological studies, offering unprecedented control of the chemical and physical environment of cells grown in microdevices, the possibility of multiplexing assays, as well as the capacity to monitor the behavior of single cells in real time while dynamically manipulating their growth medium. However, the properties of the materials employed for the fabrication of microchips that are compatible with live-cell imaging has limited the use of these techniques for a broad range of experiments. In particular, the strong absorption of a large panel of small molecules by these materials prevents the accurate delivery of compounds of interest. Here we describe a novel microsystem dedicated to live-cell imaging that (1) uses alternative materials devoid of absorptive properties, and (2) allows for dynamic in-chip control of sample temperature. Based on a proof-of-concept design that we have routinely used with non-adherent fission yeast cells, this chapter details all the steps for the fabrication and utilization of these microdevices.

1 INTRODUCTION

1.1 CONTROL OF THE CELLULAR ENVIRONMENT DURING LIVE-CELL IMAGING EXPERIMENTS

Controlling the cellular environment during live-cell imaging experiments is a powerful approach for deciphering dynamic biological phenomena at the single-cell level. Monitoring the behavior of single cells while finely modulating the composition of the culture medium, exposing cells to accurate concentrations of various compounds with high temporal resolution and dynamically regulating the temperature of these samples is becoming an integral aspect of cell biological research. Microfluidic technologies provide unprecedented and unrivaled control over these multiple parameters. However, many features of standard microfluidic chips dedicated to live-cell imaging still represent important obstacles to fully exploit these methodologies (Berthier, Young, & Beebe, 2012; Halldorsson, Lucumi, Gomez-Sjoberg, & Fleming, 2015; Paguirigan & Beebe, 2008; Regehr et al., 2009). Among these drawbacks is the well-documented absorption of a number of chemical compounds by PDMS (polydimethylsiloxane), the most widely utilized material in microfabrication (Toepke & Beebe, 2006; Wang, Douville, Takayama, & ElSayed, 2012). This particularly affects small hydrophobic molecules, a very common class of drugs used in biological studies. Here we present a versatile microfluidic platform that circumvents these issues, allowing for dynamic and precise modulation of the culture medium with a high temporal resolution as well as in-chip regulation of the sample temperature (Chen et al., 2016). This represents the only microfluidic system dedicated to live-cell imaging to date that does not interfere with the use of small molecules, offering robust and controlled compound delivery to cells under microscopic observation. We describe all the elements of the design, fabrication, and usage of this system, including a number of considerations that are critical for each of these steps. This provides a baseline for generating various types of systems that are adapted to the specific needs of different biological questions and experimental assays.

1.2 OVERVIEW OF THE MICRODEVICE AND ITS FABRICATION METHOD

The microsystem described in this chapter provides a solution for controlling the cellular microenvironment during live-cell imaging experiments (Chen et al., 2016). In particular, it allows the user to precisely and dynamically modulate the concentrations of chemical compounds of interest to which cells are exposed. To this end, this system is PDMS-free and takes advantage of the properties of thermoplastic materials that do not absorb small molecules. The device presented here is a multi-level system that integrates a microfluidic network for culturing live cells, a layer for in-chip temperature control, and a manifold for the connection of the system to flow controllers (Fig. 1A). While we detail the full protocol for building this system, simpler alternative chips can easily be made by using only parts of the device.

Our chip is based on a microfluidic network generated by hot embossing of a thin sheet of cyclic olefin copolymer (COC), a thermoplastic that has frequently been used in biomedical applications and microfluidic systems (Nunes, Ohlsson, Ordeig, & Kutter, 2010). This layer is bonded to a manifold fabricated in poly(methyl methacrylate) (PMMA) by a double-sided adhesive, which is pre-cut to integrate an additional set of microchannels that serve as an in-chip temperature control system. The whole device is then bonded to a microscopy-grade glass coverslip using paraffin wax as a sealant. To this end, melted paraffin wax is applied to the edges of the chip and allowed to spread by capillarity between the glass coverslip and the COC layer (Fig. 1A and B). Importantly, microchannels of a minimum dimension act as barriers

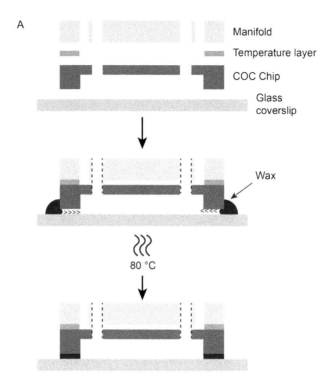

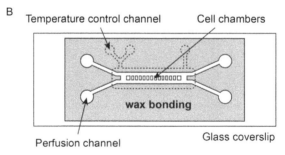

FIG. 1

Overview of the microfluidic chip fabrication. (A) The chip is a multilayered microsystem integrating a PMMA manifold for easy connections, an in-chip temperature control layer engineered in double-sided adhesive and a cell culture chamber made of COC. The assembled chip is bonded to a microscopy-grade glass coverslip using melted paraffin wax that spreads by capillarity. (B) The wax flow is stopped by the presence of the microchannels, allowing for robust bonding of the chip while keeping the channels unobstructed.

for the wax flow, permitting the sealing of the chip while preventing obstruction of the fluidic network (Fig. 1B). This chapter takes as an example a network design that we have routinely used for controlling the delivery of small molecules to a population of fission yeast cells while monitoring their response in real time by high-resolution microscopy (Fig. 2). However, applying the guidelines detailed in the following sections allows the described methods of chip fabrication and usage to be applied to a broad range of designs.

2 FABRICATION OF THE MICROFLUIDIC CHIP
2.1 DESIGN OF THE MICROFLUIDIC CHANNELS
2.1.1 Materials
The design of the microfluidic channels is a critical aspect of this procedure, as the bonding of COC microfluidic chips to glass coverslips using paraffin wax is associated with certain geometric restrictions in channel structure (see below) (Chen et al., 2016). The microfluidic network can be designed in standard computer-aided design (CAD) programs (e.g., Clewin, L-Edit, AutoCad, Corel Draw, Adobe Illustrator).

2.1.2 Considerations for the microfluidic designs
The complete COC/wax microfluidic chip is a multilayered device integrating cell culture chambers and a microfluidic temperature control device. This architecture imposes some limits to the designs that are critical for the reliable fabrication, assembly and usage of the chips. The most important aspect of the network design is the dimension of the channels. The wax bonding process relies on the spreading of melted wax between the COC microchip and the glass coverslip, with channels acting as barriers for wax flow. Generally, 40 µm represents a safe lower limit for both channel width and height, allowing robust bonding without obstruction of the network by paraffin wax during the mounting procedure (Chen et al., 2016). Systems integrating microstructures of smaller dimensions can nevertheless be fabricated, but their assembly has a higher failure rate due to wax flowing across the borders and filling the channels. Furthermore, small or isolated areas between structures may be difficult to seal: such designs should generally be avoided, and the architecture of the system should favor sealing surfaces that are easily accessible to the paraffin wax. These considerations are particularly critical when preparing designs for coupling non-adherent cells with the maintenance of a constant flow of medium throughout the experiment, as the use of such cells imposes an additional level of restrictions compared to adherent cells.

Next, when integrating a microfluidic temperature control system within the microchip (Chen et al., 2016; Velve-Casquillas et al., 2011), the positions and sizes of the observation chambers must be considered. This temperature regulation system relies on the injection of a thermalized fluid in a large chamber positioned above the cell culture channels, allowing for temperature control by heat transfer. However, this device must be calibrated to account for the heat losses that occur throughout the

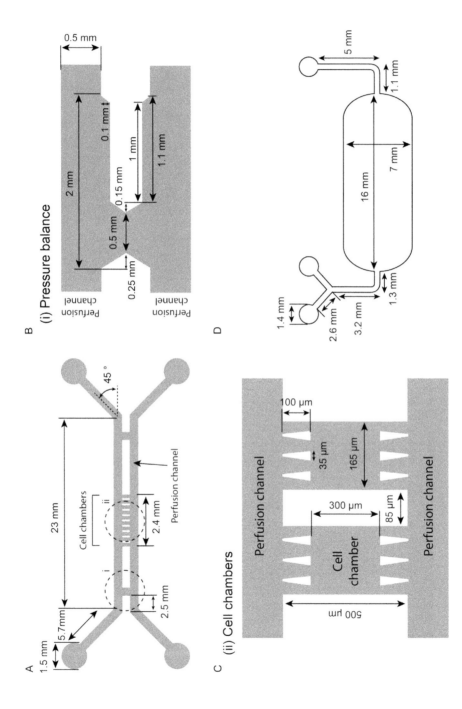

FIG. 2

Example of designs for the cell culture and temperature layers. (A) Schematic of the cell culture microfluidic network taken as an example in this chapter. We have routinely used this design to expose non-adherent fission yeast cells grown in the cell chambers to varying concentrations of different compounds. The heights of the perfusion channels and cell chambers are 100 μm and 5 μm, respectively. (i) pressure balance, see (B); (ii) cell chambers, see (C). (B) Detailed view of the pressure balance bridge positioned between the main channels. (C) Detailed view of the cell chambers. (D) Schematic of the temperature control layer. The width of the input and output channels is 0.5 mm. Our system has two inputs and one output: this allows for a rapid switch in temperature by using a second input connected to a controller set to a different temperature. (A–D) Schematics are not to scale.

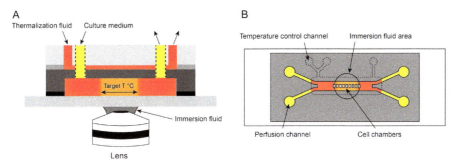

FIG. 3

Operation of the in-chip temperature control system. (A) Injection of a thermalization fluid in the top layer allows for control of the medium temperature in the cell chambers. While the medium is injected at room temperature, this has no impact on the temperature of the sample as the flow rate of the thermalization fluid (~5 mL/min) is much higher than that of the medium. However, if the lens of the microscope is in contact with the system through an immersion fluid, it acts as a heat sink and alters the temperature of the area in contact with the fluid drop. This is taken into account through a precise calibration of the system (see Fig. 6). (B) Top view of the chip. Dashed line represents the area covered by the temperature control channel. The circle corresponds to the area of the immersion fluid drop.

microfluidic network and through the immersion lens of the microscope, if one is being used (see Section 3.3.1). In this case, the lens is in contact with the glass coverslip through the immersion fluid and acts as a heat sink (Fig. 3A). Therefore, accurate sample temperature is only achieved within the area of the chip that is covered by the fluid drop: outside of this zone, channels are not subjected to the lens-mediated heat losses and are at a different temperature (Fig. 3A). Thus, all observation chambers must be positioned within a restricted area that lies within the perimeter of the fluid drop (Fig. 3B). While we generally keep all chambers in an area of 8 mm in diameter, the dimensions of this surface are easily determined and should be performed for each setup.

Attention must also be paid to the positioning of the input and output ports to ensure easy and robust fabrication of the chip manifold, to prevent overlaps between the ports of the temperature control system and those of the cell culture chambers, and to avoid putting pressure on particularly sensitive areas of the chip. For instance, although COC is a hard thermoplastic, thin COC sheets can deform under pressure. This can generate fluid displacements within the cell channels that result in movement of the cells under observation, particularly when using non-adherent cells. In addition, to allow for tight bonding around the inputs and outputs, all ports are positioned at least ~4 mm away from the edges of the chip.

Finally, it is important to note that while the flow of melted wax during the mounting procedure cannot reach structures that are positioned in open areas, it is still possible to fabricate devices containing such features. We have observed that the small space between unbound isolated COC chambers and the glass coverslip is compatible with the monitoring of non-adherent cells without major cell movements, even when a constant flow of medium is applied through communicating channels. However, if separate assays on a single chip are essential, the design must ensure proper independent sealing of each structure.

In this chapter, we present as an example the fabrication and assembly of a standard design that we routinely use with fission yeast cells (Fig. 2). This also provides a reference microfluidic network for the use of non-adherent cells.

2.2 FABRICATION OF THE COC LAYER

2.2.1 Materials

The generation of COC chips involves a first step of mold fabrication followed by a second step in which the molds are used for hot embossing of COC sheets.

For the first step, standard photolithography materials, reagents and techniques are necessary and have been extensively described elsewhere (McDonald & Whitesides, 2002; Mitra & Chakraborty, 2016; Qin, Xia, & Whitesides, 2010). Briefly, these include:

- SU-8 photoresists (e.g., Micro-Chem, USA)
- Silicon wafers (e.g., SIL-Tronix, France)
- Spincoater (e.g., WS-650 series, Laurell Technologies, USA)
- Precision hot plates (e.g., 1000-1, Electronic Micro System Ltd., UK)
- Photomasks: Film photomasks are sufficient in this context, as they offer resolutions down to ∼5–10 μm, which are below the limits recommended for the wax bonding protocol. For instance, we have routinely used Grade 4 film photomasks from JD Phototools for preparing the devices described here.
- Exposure-masking system (e.g., UV KUB2, Kloe, France). If multi-level structures are required, as in the case of the design presented in this chapter, a system with alignment capacities should be used, e.g., UV KUB3, Kloe, France.
- Photoresist developer (e.g., SU-8 developer, Micro-Chem, USA)
- PDMS (e.g., Sylgard 184, Dow Corning, USA)
- Isopropanol
- Vacuum chamber
- Oven (e.g., UNE-200, Memmert GmbH, Germany; temperatures up to 200 °C are used)

For the hot embossing procedure, the following materials are required:

- PDMS mold (see Section 2.2.2.1 for fabrication)
- Thin COC sheets of precise thickness (e.g., Topas 5013 254 μm sheets, Topas Advanced Polymers Inc., USA)
- Hot press that applies specified pressures for embossing (e.g., Atlas Manual 15Ton hydraulic press equipped with Atlas heating plates, SPECAC Inc., USA)
- Two 10 mm-thick Pyrex discs (e.g., Ref. 961428-76, Béné Inox, France)

2.2.2 *Methods*
2.2.2.1 Mold fabrication

Three separate molds are required for this procedure. A first mold in SU-8 photoresist, a second negative PDMS mold (referred to as Master 1, or M1) made using the SU-8 template, and a third positive PDMS mold (referred to as Master 2, or M2) obtained via the M1. The M2 is then used as a template for hot embossing microstructures in COC chips (Fig. 4A).

The fabrication of the SU-8 mold follows standard photolithography procedures (McDonald & Whitesides, 2002; Mitra & Chakraborty, 2016; Qin et al., 2010), which will not be detailed here. In summary, a thin layer of SU-8 photoresist is deposited on a silicon wafer with a spincoater, using specific spinning parameters to precisely achieve the target thickness depending on the heights of the microchannels. The wafer is then baked on high precision hot plates at 65°C and then 95°C. Subsequently, a photomask is used to generate the microstructures using a 365 nm exposure-masking system. In the case of a multi-level design, a second layer of SU-8 is applied following the same procedure, and alignment marks must be

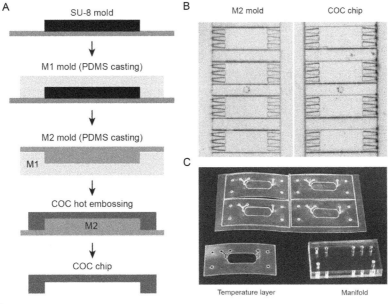

FIG. 4

Fabrication of the chip. (A) Schematic of the different steps for generating a COC chip from an initial SU-8 mold. (B) Pictures of the cell chambers in the PDMS M2 mold (left) and COC chip (right) showing the high quality of the hot embossing procedure and the accurate reproduction of the structures. The dimensions are as in Fig. 2C. Negative images are displayed for ease of visualization. (C) Picture of the pre-cut temperature control layer (top: serial production by CO_2 laser cutting; bottom left: finished temperature layer) and PMMA manifold (bottom right).

included in the photomask design in order to properly position the second photo-mask for the additional round of UV exposure. The insulated wafer is post-baked at 65 °C and then 95 °C, allowed to cool, and the structures are chemically developed. The SU-8 mold is then rinsed with isopropanol and air dried.

From the SU-8 mold, the protocol for generating the M2 mold for hot embossing is as follows:

(1) Mix PDMS with its curing agent (10:1 w/w) and degas the mixture under vacuum.
(2) Pour the liquid PDMS mixture on the SU-8 mold.
(3) Cure for 2 h at 70 °C in an oven.
(4) Peel off the negative M1 mold and bake for 12 h at 150° C.
(5) Repeat steps 1–5 using this M1 mold as a template, but with a thinner coat (~3 mm) of 5:1 (w/w) PDMS (no treatment of the M1 mold is required).
(6) The obtained positive M2 mold is allowed to further harden for a minimum of 2 h at 200 °C. The reduced thickness of M2 is important to limiting deformation of the substrate during the hot embossing process.

At this point, M2 is ready to be used for generating COC microchips. Note that such a mold can be exploited for many rounds (>100) of hot embossing without any particular treatment.

2.2.2.2 Hot embossing of COC chips

COC is cut from a sheet of precise thickness and placed onto the M2 mold (we routinely use a piece of COC that extends ~5 mm beyond the outermost edges of the design). The COC/M2 montage is then sandwiched between two 10 mm-thick Pyrex discs and inserted between the heating plates of the hot press under minimal pressure. It is critical to keep the M2 mold above the COC, as degassing of the PDMS during the procedure can generate bubbles in the COC chip if the system is placed in the inverse configuration. To obtain thin COC chips that faithfully reproduce the mold microstructures (Fig. 4B), the following procedure is used:

(1) Heat the press to 180 °C.
(2) Apply a pressure of 200 kg for 20 s.
(3) Set the temperature to 70 °C and let the system slowly cool down.
(4) Release the pressure.
(5) Peel the COC chip off the M2 mold.
(6) If necessary, the quality and size of the structures can be assessed using a profilometer.

2.3 FABRICATION OF THE IN-CHIP TEMPERATURE CONTROL LAYER

2.3.1 Considerations for the integration of an in-chip temperature control system

Controlling the temperature of the cellular environment is critical when performing live-cell imaging experiments. Many systems exist for regulating the temperature of samples under microscopic observation, from objective heaters to large environmental chambers to an in-chip microfluidic temperature control device. Here we present the integration of a microfluidic system for temperature regulation (Velve-Casquillas et al., 2011) made from double-sided adhesive (Chen et al., 2016). However, as this adds complexity to the fabrication and usage of the chips, it is important to consider whether such a layer is critical. Indeed, for maintaining a constant temperature above ambient, incubation chambers or objective heaters may be sufficient for providing temperature regulation that is accurate enough for most live-cell imaging assays. In this case, the design of the microfluidic network is not constrained by the in-chip temperature control system (see previous sections). In contrast, when cooling the sample below the ambient or if it is necessary to perform rapid temperature shifts, the microfluidic layer that we have added to our microsystem is an interesting solution, despite its added complexity.

2.3.2 Materials
- Double-sided adhesive of precise thickness (e.g., 81 μm-thick ARcare 90880 medical tape, Adhesive Research Inc., USA)
- Cutting the adhesive thermalization chamber can be achieved using simple razor blades. However, for normalizing the production of the chips and fabricating several identical systems at a time, we are using a CO_2 laser cutter (Speedy 100, 60W, Trotec, Austria).

2.3.3 Design and methods
We have implemented a large thermalization chamber of 7×16 mm that is then positioned above the cell culture chambers (Fig. 2D). The dimensions of this chamber are relatively flexible, but to ensure robust temperature regulation, it must accommodate a $\sim$5 mL/min flow rate of the thermalization fluid that is pumped through the chamber by the temperature controller (Cherry Biotech, France). The total dimensions of this layer must be adapted to the size of the design and the glass coverslip that will be used to generate the chip.

When using a CO_2 laser cutter, the design can be produced with a standard CAD program (e.g., Adobe Illustrator, Corel Draw). The cutting parameters must then be optimized depending on the laser cutter to generate accurate and non-deformed patterns (e.g., Power 45%, Speed 7%, PPI/Hz 1000 using a Speedy 100, 60 W system from Trotec, Austria). The presence of liners on both faces of the adhesive protects the surface and allows for the storage of pre-cut temperature layers for long periods of time (Fig. 4C).

2.4 FABRICATION OF A CONNECTING MANIFOLD
While not fundamental to the operation of the chip, the addition of an adapted manifold plays an important role in facilitating the connection of the input and output tubings as well as improving the reliability of the system for applying a flow of medium during long-term experiments.

2.4.1 Materials
- 6 mm-thick extruded PMMA plates (e.g. from Weber Metaux, France)
- While precision cutting and drilling machines can be used for the fabrication of the PMMA manifold, we have obtained higher quality parts using a CO_2 laser cutter (Speedy 100, 60W, Trotec, Austria). This also allows for the simultaneous production of multiple pieces.
- Scalpel
- Isopropanol
- Pressurized air

2.4.2 Design and methods
The dimensions of the manifold must be adapted to the size of the glass coverslips on which the system will be mounted. For 24×60 mm microscopy-grade coverslips, we fabricate 20×40 mm manifolds and temperature control layers. All connection ports are 1.5 mm in diameter to accommodate standard 1/16″ tubings.

For cutting the manifold using a CO_2 laser cutter, a similar approach as for the temperature control layer is applied, but the parameters have to be adjusted (for instance, Power 76%, Speed 0.35%, PPI/Hz 1000 using a Speedy 100, 60W system from Trotec, Austria).

Note that the edges of the manifold may be slightly deformed, notably due to the heat generated by the laser. This can produce structures that prevent reliable bonding of the thermalization layer and affect the flatness of the assembly, resulting in poor bonding to the glass coverslip by wax (see Section 2.6). To avoid this issue, the edges of the manifold are systematically scraped using a scalpel. The PMMA manifold is then cleaned with isopropanol and dried using pressurized air (Fig. 4C).

2.5 ASSEMBLY OF THE COMPLETE MICROFLUIDIC CHIP
2.5.1 Materials
- Razor blade
- Scalpel
- Needle (e.g., 23 G blunt-end safety tip needles)
- Gas burner
- Isopropanol
- Pressurized air

2.5.2 Methods
2.5.2.1 Step 1: Assembly of the temperature control layer
(1) Peel off the protective liner from one side of the pre-cut tape (see Section 3.3).
(2) Bond the temperature control layer to the PMMA manifold with manual pressure, paying particular attention to the alignment of the connecting ports (Fig. 5A and B).

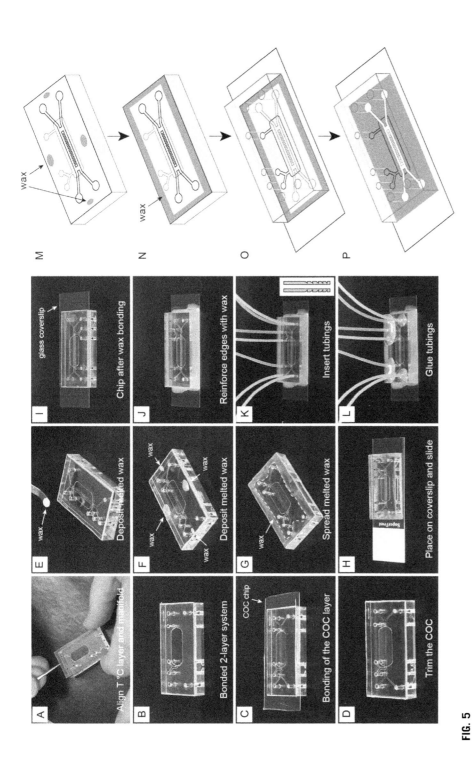

FIG. 5

Bonding of the COC chip to a glass coverslip using paraffin wax as a sealant. (A–L) Pictures of the different steps of the protocol. Between (F) and (G), the hardened drops of wax are spread along the edges of the chip using a heated spatula (dashed line in G delineates the inner front of the spread wax area). The inset in (K) is a schematic of the small indents that are made in the tubings prior to inserting them in the connecting ports in order to reinforce their final bonding. (M), (N), (O), and (P) are schematics corresponding to the pictures in (F), (G), (H), and (I), respectively.

2.5.2.2 Step 2: Mounting of the COC microfluidic network

(1) Peel off the second protective liner of the temperature control layer.

(2) Bond the thin, hot-embossed COC layer to the tape with manual pressure, paying particular attention to the alignment of the connecting ports (Fig. 5C). To avoid breaking the COC sheet, do not apply pressure in the area covered by the temperature control system.

(3) Remove excess COC using a standard razor blade (Fig. 5D). The final COC layer has the same dimensions as the manifold and temperature control layers. As for the laser cutting of the PMMA manifold, this generates small deformations on the edges of the chip, preventing high quality wax-mediated bonding of the chip to a glass coverslip. To prevent this, the edges of the mounted COC layer are scraped using a scalpel.

(4) Re-apply manual pressure to the COC layer to ensure robust bonding.

(5) The ports for medium flow are then opened using a needle that has been heated with a gas burner. To prevent any deformation that would alter wax bonding during this procedure, holes are always made from the COC side in the direction of the PMMA manifold. A new, unheated needle is then used to clean the ports.

(6) Apply a final manual pressure to the COC layer.

(7) Clean the exposed COC surface with isopropanol and dry with pressurized air.

The fully assembled chip is ready to be bonded to a glass coverslip.

2.6 BONDING OF THE MICROCHIP TO A GLASS COVERSLIP AND ASSEMBLY OF THE CONNECTIONS

This is the most critical part of the procedure to generate reliable COC/wax chips, as it ensures the sealing of the COC to a microscopy-grade glass coverslip (Chen et al., 2016).

2.6.1 Materials

- Microscopy-grade glass coverslip of the adapted size (e.g., $0.17 \times 24 \times 60$ mm)
- Microscope glass slide (e.g., 26×76 mm)
- Paraffin wax (e.g., Ref. 411663, Sigma Aldrich, USA)
- Chattaway spatula
- Hot plate (e.g., C-Mag HS 7, IKA, USA)
- Weight ~ 1200 g (we routinely use boxes of microscope slides)
- Isopropanol
- Pressurized air
- Connecting tubing (e.g., PTFE $1/16''$ OD $\times 0.5$ mm ID for the medium and PTFE $1/16''$ OD $\times 1$ mm ID for the thermalization fluid, IDEX, USA)
- Scalpel
- Epoxy glue (e.g., SADER, Bostik, USA)

2.6.2 Methods

2.6.2.1 Step 1: Wax bonding of the microsystem

(1) Melt $\sim$5 mg (1/2 pellet) of paraffin wax on a spatula using a gas burner and apply a thin coat of melted wax along the long edges of the COC chip, covering a $\sim$3 mm-wide area. The thickness of the wax layer should be as homogenous as possible (Fig. 5E–G, M, and N). If necessary, the wax layer can be extended to areas that need to be fully bonded. Allow the wax to harden at room temperature.

(2) Repeat the same procedure along the short edges of the COC chip using $\sim$2.5 mg (1/4 pellet) of wax to cover a $\sim$2 mm-wide area (Fig. 5E–G, M, and N).

(3) Clean the glass coverslip with isopropanol, dry with pressurized air, and place the coverslip on a standard microscope slide.

(4) Position the microsystem on the glass coverslip and put the entire setup on a hot plate at room temperature (Fig. 5H and O).

(5) Apply a weight of $\sim$1200 g at the top of the assembled system. It is important to note that the layer of wax slightly alters the heights of the channels. This extra thickness depends on the weight applied ($\sim$5 μm in these conditions (Chen et al., 2016)).

(6) Put a small pellet of paraffin wax on another microscope slide and place the slide on the hot plate, next to the microsystem. This will serve to monitor the melting of the wax.

(7) Turn on the hot plate to 80 °C and wait until the wax indicator melts (at this point, the melted wax in the microsystem will spread by capillary action between the COC layer and the glass coverslip; Fig. 5P).

(8) Switch off the hot plate, wait for 10 s, carefully remove the weight, and rapidly place the microsystem on a surface at room temperature. Maintaining the microsystem at high temperature for too long often results in the wax flowing through and obstructing the microfluidic channels.

(9) Allow the system to cool and the wax to harden, thus establishing a strong bonding (Fig. 5I and P).
  – *Optional*: If the wax has not spread equally and some areas of the chip are still unbound (this can notably happen if there is dust or if the amount of wax used is insufficient), proceed as follows:
  (a) Deposit additional melted wax at the appropriate edges of the chip. During this step, it is important to avoid encapsulating air bubbles. Air must be allowed to exit the system so that all areas are robustly bonded.
  (b) Heat a clean spatula and apply it to the appropriate areas of the coverslip to locally melt the wax and allow complete coverage of the surface to be bound.

(10) Complete the sealing by applying a last layer of melted wax around the COC/glass interface of the entire microsystem using a spatula (Fig. 5J).

(11) Allow the chip to cool.

2.6.2.2 Step 2: Assembly of the connections

The use of PTFE tubings is critical to prevent drug absorption. Indeed, standard tubings used in microfluidics (e.g., Tygon tubing) have significant absorptive properties.

(1) Make small indents to the outer surface of the tubings using a scalpel (Fig. 5K inset). This is only necessary on the areas of the tubings that will be inserted into the ports of the PMMA manifold and allows for more robust bonding when glued.
(2) Insert the tubings in all the ports and seal the connections using epoxy glue (Fig. 5K and L).

The chip fabrication is now complete and the system ready to be used.

3 SETTING UP AND USING THE MICROFLUIDIC SYSTEM
3.1 PRELIMINARY PROCEDURES
3.1.1 Drug absorption tests

The COC/wax microdevices have been developed to provide a solution for the well-described absorption of small molecules by PDMS, a major obstacle for a number of biological studies (Toepke & Beebe, 2006; Wang et al., 2012). We previously showed that these new chips show no absorption of the tested small molecules (Chen et al., 2016), taking advantage of the fluorescent marker Rhodamine B, which has commonly been used to test the absorptive properties of microfluidic chips (Ren, Zhao, Su, Ryan, & Wu, 2010; Roman, Hlaus, Bass, Seelhammer, & Culbertson, 2005; Sasaki, Onoe, Osaki, Kawano, & Takeuchi, 2010), as well as a much more sensitive cell-based assay (Chen et al., 2016; Coudreuse & Nurse, 2010). Nevertheless, it remains essential to validate the COC/wax devices for each drug that has not been reported to be compatible with these systems. Well-defined and highly sensitive read-outs are therefore necessary. Initial tests can be performed by exposing cells to the compounds of interest between two glass coverslips, one of which having been pre-coated with a thin layer of paraffin wax. However, this only allows for revealing major absorption issues (Chen et al., 2016): a drug that shows no absorption in this assay must still be tested in the context of the final chip, as the surface to volume ratio in a chip is much higher, thus favoring absorption phenomena.

3.1.2 Calibrating the temperature control system

As mentioned earlier, the integration of the in-chip microfluidic temperature control system offers a number of unique features, in particular the possibility of working below ambient temperature as well as inducing fast temperature switches. The latter is especially advantageous when studying temperature-sensitive processes or using temperature-sensitive mutants. To use this layer, the chip must be connected to a system that controls both the flow rate and the temperature of the injected thermalized fluid (Cherry Temp, Cherry Biotech, France). However, this system must be calibrated to account for the heat exchanges that occur with the microscope objective when using immersion lenses (Chen et al., 2016). This is an essential step for ensuring an accurate target temperature within the controlled area (see Section 2.1.2). The main feature of the chip that alters the temperature in the microsystem is the height of the cell channels and the thickness of the COC sheet above. Thus, for a given set of such parameters, the design of the fluidic network can be extensively modified without the need for a new calibration.

3.1.2.1 Materials
— Resistance Temperature Detector (RTD) on a glass coverslip. The RTD (Fig. 6A and B) is fabricated by electrodeposition as follows: 5 nm titanium (adhesion), 70 nm platinum (conduction), 300 nm silicone nitride (insulating layer), 2 nm Ag (bonding electrical wires). RTDs can be ordered from microfabrication core facilities, as high-end equipment and expertise are required for their production. The protocol for the fabrication of the RTD will therefore not be developed in this chapter.
— Data acquisition unit (e.g., model 34972A from Agilent Technologies; a MATLAB® interface is available for driving the unit). The apparatus is used according to manufacturer's instructions.
— Conductive epoxy glue (e.g., CW2400, Chemtronics, USA)
— Electrical wire
— Needles (e.g., 23 G blunt-end safety tip needles)
— Paraffin wax (e.g., Ref. 411663, Sigma Aldrich, USA)
— Parafilm
— Polystyrene box
— High precision oven

3.1.2.2 Methods
— *Step 1: Calibration of the electrode*
   (1) Connect electrical wires to the RTD electrode with conductive epoxy glue.
   (2) Mount the full COC/wax chip to be calibrated on the electrode coverslip (including the connecting tubings).
   (3) Inject water in the cell culture channels.
   (4) Close all the ports of the chip by filling needles with melted paraffin wax and wrapping them in parafilm to prevent evaporation.
   (5) Connect the wires to the data acquisition unit.
   (6) Place the system in a polystyrene box (adiabatic container) to prevent heat transfer and incubate in a high precision oven.
   (7) Wait for the temperature of the system to stabilize (>3 h).
   (8) Establish the relationship between the temperature (imposed by the oven) and the resistance of the electrode (measured). Repeat steps 7 and 8 for at least two additional temperatures (choose temperatures within the range of your experiments; typically we use temperatures between 20 °C and 40 °C). A linear correlation between temperature and resistance is then established (Fig. 6C). The electrode is calibrated.

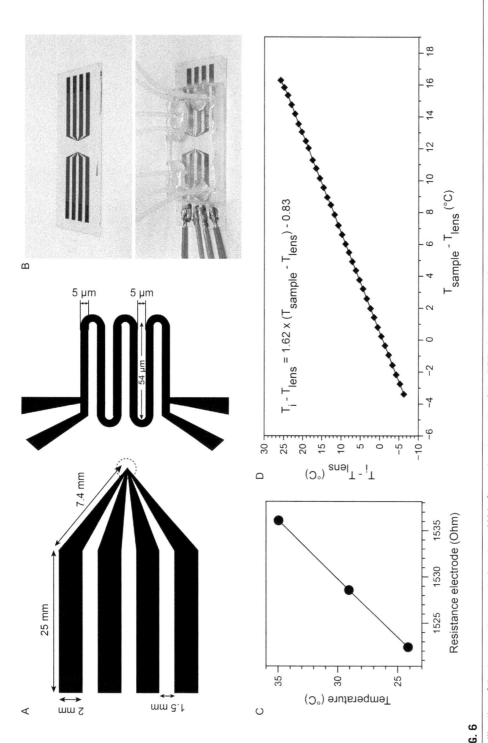

FIG. 6

Calibration of the temperature control system. (A) Left panel: schematic of the RTD electrode used for the temperature calibration. Right panel: detailed view of the electrode (zoom of the area in the dashed circle, left panel). (B) Top: picture of an RTD electrode deposited on a glass coverslip. Bottom: picture of a full chip mounted on a calibration electrode. (C) Calibration curve for the chip. This curve indicates the temperature that must be set for the thermalization fluid, provided the temperature of the lens and the target temperature of the sample (data as in Chen et al., 2016). The data provided in (C) and (D) are examples of calibration curves that were made for the microfluidic network used as an example in this chapter (see Fig. 2).

— *Step 2: Calibration of the chip*

(1) Mount the calibration chip on the microscope in normal experimental conditions (in contact with the lens through the appropriate immersion fluid, if applicable).

(2) Start the temperature control system (Cherry Temp, Cherry Biotech, France) using the non-calibrated mode.

(3) Set the system to a specific temperature, wait 2–3 min to allow for stabilization, and note the temperature of the RTD (measured through the data acquisition unit, as above). The temperature of the lens must also be taken. This can be determined by a temperature probe that is part of the Cherry Temp (the temperature is provided by its software) or by a separate probe.

(4) Repeat step 3 for a broad range of temperatures and establish the relationship $T_{inj} - T_{lens} = f(T_{RTD} - T_{lens})$, where T_{inj} is the injected temperature indicated by the software, T_{lens} is the temperature of the lens provided by the software and T_{RTD} is the temperature measured by the electrode (Fig. 6D).

(5) This calibration equation can then be implemented in the control software by the provider (Cherry Biotech, France). This allows the system to inject the thermalization fluid at the appropriate temperature for a given temperature of the lens and a given target temperature.

Note that this protocol describes the calibration of a chip used with immersion lenses. Air lenses are not in contact with the glass coverslip and therefore do not act as heat sinks. Nevertheless, a similar calibration protocol should be used in this case to account for heat losses that can occur elsewhere in the fluidic network.

3.2 CELL LOADING

Loading the cells into a microdevice is a critical part of the experiment, as particular attention must be paid to avoid generating bubbles in the microfluidic chip. To prevent this problem, sterilized water is first injected in the chip, which is incubated at room temperature overnight to allow for the removal of all bubbles in the system. The chip is then carefully washed with medium using a syringe, and cells can then be loaded using syringes with needles of a diameter adapted to the tubings. When using designs with lateral chambers such as the one presented here, some of the output ports must be temporarily blocked to force cells into the observation chambers. The chip can then be connected to the various devices that allow for the control of the cellular environment in the microsystem. This can be achieved via a large set of specific connectors that are commercially available for accommodating a broad spectrum of fluidic networks (e.g., connectors from IDEX, USA). The chip is now ready to be mounted on a microscope. Note that it is important to use a system that maintains the chip in the slide holder of the microscope (e.g., slide spring clips or adhesive), as the weight of the connections and tubings can easily lift the entire system off the stage.

3.3 ENVIRONMENTAL CONTROL IN THE MICROSYSTEM

3.3.1 Temperature control

The regulation of sample temperature in the chip when using large thermalization chambers is straightforward. One should however consider the potential effect of applying a constant flow of medium through the chip during the experiment. In this case, simply placing the bottle of medium in the thermalization chamber prevents temperature deviations. Alternatively, having a sufficiently long portion of tubing for medium delivery within the thermalized chamber is sufficient for heating the medium prior to its arrival in the microchip. An additional level of dynamic control of the temperature can be provided by the in-chip temperature control system (see Section 3.1.2). This device allows fast temperature switches as well as cooling of the sample below ambient. Interestingly, it takes into account potential variations in the temperature of the lens (for immersion systems) in real-time using data from the calibration described in Section 3.1.2.2. This controller is operated according to manufacturer's instructions.

3.3.2 Fluid control

For long-term live-cell imaging experiments, it is critical to apply a constant flow of fresh medium in the chip in order to maintain optimal growth conditions for the cells. For instance, in the absence of flow, we have observed a significant decrease in cell size at division of fission yeast cells grown in microsystems after only a few hours, a phenotype that reflects a state of nutritional stress. While connecting the chip to an elevated vial of fresh medium is sufficient to generate a flow of medium by gravity, this approach does not maintain an accurate and constant rate of medium renewal. Below is a list of alternatives with their advantages and drawbacks:

— *Precision syringe pumps*: These devices permit accurate and constant flow control in the chip (for instance, Harvard Apparatus provides a full range of syringe pump systems). However, attention must be paid to assess the total amount of medium required for the length of an entire experiment, as this volume is limited by the size of the syringe that is used. In addition, multiplexing is more difficult.

— *Pressure controllers*: These systems provide accurate and stable medium flow, and they can be connected to medium vials of any size, making them extremely versatile (e.g., OB1 Pressure controller, Elveflow, France). Furthermore, they are generally provided with control programs that allow for the implementation of automated changes in the pressure applied. However, as these systems are based on pressure control, the actual flow rates in the chip are hard to predict since they are highly dependent on the fluidic resistance in the downstream microfluidic network.

— *Flow controllers*: For biological experiments, we favor this solution, as it provides a complete and accurate control of the flow rate in the chip while keeping the advantages of pressure control systems. This approach relies on the use of flow sensors that precisely measure and maintain flow rates by adjusting the operation of pressure controllers in real-time (for example, we have used OB1 pressure controllers coupled with flow sensors from Elveflow, France).

4 DRUG DELIVERY IN THE MICROSYSTEM
4.1 CONSTANT DRUG DELIVERY VS MEDIA SWITCH

Constant drug delivery can simply be achieved by adding the drug to the medium stock. To induce media switches, two solutions are available. First, the medium stock can be manually changed by the user. This is however tedious when complex patterns of media switches are necessary, and it promotes the injection of air in the system during the switches. A more reliable solution is to use a valve matrix (e.g., MUX, Elveflow, France; used according to manufacturer's instructions), in which each valve can be connected to a different source of medium. This allows for rapid, reliable, and automated changes in the medium along with a high versatility in the exposure of cells to different drugs, individually or simultaneously, at different concentrations.

4.2 DRUG DELIVERY WITH COMPLEX DYNAMICS

While the use of a matrix of valves allows for automated switches between different media, it is not sufficient when precise and complex dynamics in drug concentrations must be applied. In addition, when the experiment requires the sequential exposure of cells to a large number of different concentrations, this approach can rapidly become laborious. To control the target drug concentration in the chip with high precision and a very high degree of freedom, we use controlled mixing of a limited number of stock solutions. As the flows of fluids in microdevices are laminar, this requires the integration of a micromixer between the media sources and the chip.

4.2.1 Micromixer fabrication

We take advantage of a standard multi-level microdevice that integrates series of herringbone staggers (Lee, Chang, Wang, & Fu, 2011; Williams, Longmuir, & Yager, 2008) (Fig. 7). We have used this design for a broad range of experiments, and the number of input and output ports is the only parameter that we have changed depending on the needs of each experiment. Importantly, to prevent the absorption of chemical compounds by the mixing device, we fabricate the systems using COC and PMMA.

4.2.1.1 Materials
— The same reagents and equipment as for the fabrication of the COC layer and PMMA manifold are required (see Sections 2.2 and 2.4)
— Plasma cleaner (e.g., PDC-002 plasma cleaner, Harrick Plasma, USA)
— Isopropanol
— Cyclohexane (e.g., Ref. 179191, Sigma Aldrich, USA)
— 10 cm-diameter glass Petri dish with glass cover
— 4-cm-diameter glass Petri dish
— Parchment paper
— Nail polish

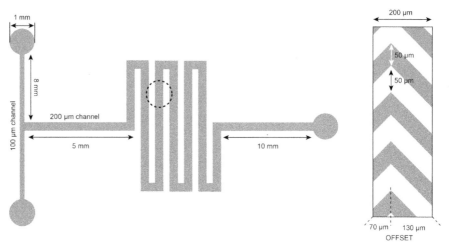

FIG. 7

Mixing media. Schematic of a micromixer. For accurate dynamic control of the medium and concentrations of compounds of interest, real-time mixing of various solutions may be necessary. Left panel: global structure of the micromixer network. The total length of the serpentine is 30 mm. The height of the channels is 150 μm. This shows a 2:1 mixer, but we have used a similar design for 3:1 and 2:2 systems. Right panel: detailed view of the herringbone staggers fabricated on the ceiling of the channels in the serpentine zone to promote fluid mixing (dashed circle in the left panel). A total of 200 staggers are used. Every 10 structures, the offset (70 μm/130 μm) is inverted. The angle of the structures is 45 degrees.

4.2.1.2 Methods

(1) Fabricate a PMMA manifold following the procedure for the fabrication of the COC chip described in Section 2.4.

(2) Fabricate a COC layer with the micromixer structures. The protocol is identical to that employed for the COC chip (Section 2.2), except that two pieces of COC are used for the hot embossing of the micromixer, generating a thicker layer (alternatively, a thicker initial sheet of COC can be used). This allows for a more resistant system.

(3) Clean the surfaces of the COC and PMMA parts with isopropanol and air dry.

(4) Place both parts with surfaces to be bonded facing upward in a plasma cleaner and treat for 30 min at high power.

(5) Transfer the parts into a 10-cm glass Petri dish, still with the bonding surfaces facing upward.

(6) Place a 4-cm glass Petri dish in the large dish together with the COC and PMMA parts, and add ~250 μL of cyclohexane in this small dish.

(7) Cover the glass dish and allow the cyclohexane to evaporate for 20 min. This vapor treatment step allows for mild dissolution of the surfaces of the parts without damaging the structures.

(8) Align the two parts (micromixer structures facing the PMMA block) and apply a mild manual pressure to generate a fully closed micromixer.

(9) Place the micromixer between two Pyrex disks in the same hot press as for the fabrication of the COC chip. To avoid any bonding of the materials to the Pyrex, intercalate parchment paper between the Pyrex disks and the micromixer.

(10) Heat the press to 110°C and apply a pressure of 200 kg for 30 min.

(11) Reduce the press temperature to 40 °C.

(12) Release the pressure.

(13) Remove the micromixer from the press and apply a thin coat of nail polish to the edges of the micromixer to ensure strong sealing.

At this point the micromixer is ready for use, and tubings can be inserted in the system following the same procedure as for the COC/wax chip. As above, PTFE or similar tubings must be used to prevent drug absorption. This micromixer can be used for several independent experiments if it is thoroughly washed with ethanol and sterile water, or alternatively with a diluted bleach solution.

4.2.2 Mixing media

The micromixer can then be coupled with flow sensors for each input. Each individual flow sensor is in turn connected to a valve that allows for the distribution of medium containing a given concentration of a compound of interest (Fig. 8A). By simply varying the relative flow rates of the different inputs, any concentration can be obtained from a limited number of stock solutions. Importantly, the very high time resolution of the flow sensor-based fluid control system allows for tight dynamic control of the concentration of drugs to which cells are exposed (Fig. 8B).

5 GENERAL CONCLUSIONS: PROS AND CONS

Controlling the cellular environment during live-cell imaging experiments has become a key aspect of cell biological studies. Microfluidic technologies provide a novel means to achieve dynamic modulation of a host of parameters with high time resolution while monitoring the response of single cells in real-time. However, drug delivery has remained a challenge due to the absorptive properties of the most commonly used material in microfabrication for biology, PDMS. Here we provide a method for fabricating PDMS-free microsystems that circumvent this issue. The proposed strategy has a few drawbacks: it requires access to specific equipment and some expertise in microfabrication. In addition, although these chips are fully compatible with high quality fluorescence microscopy, they only permit a basic level of transmission microscopy, as is the case for most microsystems that combine microfluidics with live-cell imaging. However, despite some design restrictions that are intrinsic to the wax bonding protocol, this method remains very versatile and can be adapted for a broad range of designs, model organisms, and approaches.

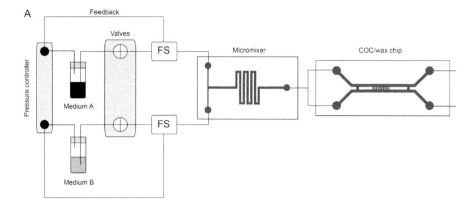

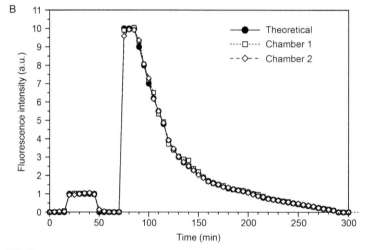

FIG. 8

Dynamic control of the medium in the microsystem. (A) Example of the full setup for modulating the cellular environment in real-time. A simple switch from one medium to another can be achieved using valves. For more complex control, the ratio of mixing between the two different media is insured by separate flow sensors that control the operation of a pressure controller. This allows for a dynamic change in medium composition. (B) Example of dynamic control of the concentration of Rhodamine B in the cell chambers of the design used as an example in this chapter. Three flow sensors connected to 0, 100, and 200 μM Rhodamine B were used together with a 3:1 micromixer. A first round of 15 min at 0 μM was performed to determine the background fluorescence in the chambers, followed by 15 min at 10 μM (the maximum concentration used in this experiment) to obtain the maximum intensity. Subsequently, the dynamic sequence was started. While the total flow rate was maintained at 40 μL/min (20 μL/min in each feeding channel), the individual flow rates of each of the solutions were established according to the theoretical curve. Images were then acquired every 5 min using a spinning disk confocal microscope for two independent chambers (the first and last chambers of the row). Average fluorescence in each chamber was determined, the background was subtracted and the signals were normalized to the maximum fluorescence at 10 μM Rhodamine B (see above). Note that this experiment was performed in the presence of fission yeast cells, which were used as reference points for the focal plane.

Importantly, while this property should be tested for any new compound, we have not observed absorption of any of the drugs that we have assessed. Finally, coupled with the in-chip temperature control system, upstream flow controllers, valve matrices, and custom-made micromixers, virtually all parameters of the cellular environment can be automatically and dynamically altered, independently or simultaneously, allowing the user to perform extremely complex assays. This platform is, to our knowledge, the only system dedicated to live-cell imaging to date that is fully compatible with small molecules and that offers such a complete control of the cellular environment.

ACKNOWLEDGMENTS

We thank members of the SyntheCell and Genome Duplication and Maintenance teams for fruitful discussions. We are grateful to Michael O'Brien from Adhesives Research Ireland for advice on the integration of adhesives in the chips, Cherry Biotech for helping with the integration of the temperature control system and Olivier de Sagazan from the NanoRennes core facility for the fabrication of metal electrodes. We thank Pei-Yun Jenny Wu for critically reading the chapter. J.M.-G. was supported by funding to D.C. from the European Research Council under the European Union's Seventh Framework Programme (FP7/2007-2013)/ERC Grant Agreement no. 310849. J.B. was funded by a research grant to D.C. from the Direction Générale des Armées (DGA). This work was also supported by a collaborative grant from the Région Bretagne, France.

REFERENCES

Berthier, E., Young, E. W. K., & Beebe, D. (2012). Engineers are from PDMS-land, biologists are from Polystyrenia. *Lab on a Chip*, *12*, 1224–1237.

Chen, T., Gómez-Escoda, B., Munoz-Garcia, J., Babic, J., Griscom, L., Wu, P.-Y. J., et al. (2016). A drug-compatible and temperature-controlled microfluidic device for live-cell imaging. *Open Biology*, *6*, 160156.

Coudreuse, D., & Nurse, P. (2010). Driving the cell cycle with a minimal CDK control network. *Nature*, *468*, 1074–1079.

Halldorsson, S., Lucumi, E., Gomez-Sjoberg, R., & Fleming, R. M. T. (2015). Advantages and challenges of microfluidic cell culture in polydimethylsiloxane devices. *Biosensors & Bioelectronics*, *63*, 218–231.

Lee, C.-Y., Chang, C.-L., Wang, Y.-N., & Fu, L.-M. (2011). Microfluidic mixing: A review. *International Journal of Molecular Sciences*, *12*, 3263–3287.

McDonald, J. C., & Whitesides, G. M. (2002). Poly(dimethylsiloxane) as a material for fabricating microfluidic devices. *Accounts of Chemical Research*, *35*, 491–499.

Mitra, S. K., & Chakraborty, S. (2016). *Microfluidics and nanofluidics handbook.* CRC Press.

Nunes, P. S., Ohlsson, P. D., Ordeig, O., & Kutter, J. P. (2010). Cyclic olefin polymers: Emerging materials for lab-on-a-chip applications. *Microfluidics and Nanofluidics*, *9*, 145–161.

Paguirigan, A. L., & Beebe, D. J. (2008). Microfluidics meet cell biology: Bridging the gap by validation and application of microscale techniques for cell biological assays. *BioEssays*, *30*, 811–821.

Qin, D., Xia, Y., & Whitesides, G. M. (2010). Soft lithography for micro- and nanoscale patterning. *Nature Protocols, 5,* 491–502.

Regehr, K. J., Domenech, M., Koepsel, J. T., Carver, K. C., Ellison-Zelski, S. J., Murphy, W. L., et al. (2009). Biological implications of polydimethylsiloxane-based microfluidic cell culture. *Lab on a Chip, 9,* 2132–2139.

Ren, K., Zhao, Y., Su, J., Ryan, D., & Wu, H. (2010). Convenient method for modifying poly(dimethylsiloxane) to be airtight and resistive against absorption of small molecules. *Analytical Chemistry, 82,* 5965–5971.

Roman, G. T., Hlaus, T., Bass, K. J., Seelhammer, T. G., & Culbertson, C. T. (2005). Sol-gel modified poly(dimethylsiloxane) microfluidic devices with high electroosmotic mobilities and hydrophilic channel wall characteristics. *Analytical Chemistry, 77,* 1414–1422.

Sasaki, H., Onoe, H., Osaki, T., Kawano, R., & Takeuchi, S. (2010). Parylene-coating in PDMS microfluidic channels prevents the absorption of fluorescent dyes. *Sensors and Actuators B: Chemical, 150,* 478–482.

Toepke, M. W., & Beebe, D. J. (2006). PDMS absorption of small molecules and consequences in microfluidic applications. *Lab on a Chip, 6,* 1484–1486.

Velve-Casquillas, G., Fu, C., Le Berre, M., Cramer, J., Meance, S., Plecis, A., et al. (2011). Fast microfluidic temperature control for high resolution live cell imaging. *Lab on a Chip, 11,* 484–489.

Wang, J. D., Douville, N. J., Takayama, S., & ElSayed, M. (2012). Quantitative analysis of molecular absorption into PDMS microfluidic channels. *Annals of Biomedical Engineering, 40,* 1862–1873.

Williams, M. S., Longmuir, K. J., & Yager, P. (2008). A practical guide to the staggered herringbone mixer. *Lab on a Chip, 8,* 1121–1129.

Controllable stress patterns over multi-generation timescale in microfluidic devices

Youlian Goulev[*,†,‡,§,1], **Audrey Matifas**[*,†,‡,§], **Gilles Charvin**[*,†,‡,§,1]

Department of Developmental Biology and Stem Cells, Institut de Génétique et de Biologie Moléculaire et Cellulaire, Illkirch, France
[†]*Centre National de la Recherche Scientifique, UMR7104, Illkirch, France*
[‡]*Institut National de la Santé et de la Recherche Médicale, Illkirch, France*
[§]*Université de Strasbourg, Illkirch, France*
[1]*Corresponding authors: e-mail address: youlian.goulev@gmail.com; charvin@igbmc.fr*

CHAPTER OUTLINE

Abstract

The generation of complex temporal stress patterns may be instrumental to investigate the adaptive properties of individual cells submitted to environmental stress on physiological timescale. However, it is difficult to accurately control stress concentration over time in bulk experiments. Here, we describe a microfluidics-based protocol to induce tightly controllable H_2O_2 stress in budding yeast while constantly monitoring cell growth with single cell resolution over multi-generation timescale. Moreover, we describe a simple methodology to produce ramping H_2O_2 stress to investigate the homeostatic properties of the H_2O_2 scavenging system.

1 INTRODUCTION

Environmental changes (e.g., temperature, chemical composition of the medium, osmotic pressure, pH, desiccation) are a major threat for living organisms. To prevent damages associated with these stresses, unicellular organisms have evolved complex physiological responses. Genetic studies have successfully identified a vast number of genes involved in stress response and stress transduction pathways and have been extensively characterized at the biochemical level (Auesukaree, 2017; Hohmann, 2002; Morano, Grant, & Moye-Rowley, 2012). However, some functional properties of stress response, such as the specificity of the response, the accuracy of adaptation, the acquisition of tolerance or the origin of cross-protection, remain to be elucidated (Davies, 2016; Guan, Haroon, Bravo, Will, & Gasch, 2012). This requires the development of integrative approaches bridging classical genetics, quantitative monitoring of cellular response, with an accurate real-time control of environmental conditions (Castillo-Hair, Igoshin, & Tabor, 2015; Muzzey & van Oudenaarden, 2009). For instance, previous studies have investigated the cellular response to periodic (Hersen et al., 2008; Mettetal et al., 2008) or ramping (Muzzey & van Oudenaarden, 2009; Young, Locke, & Elowitz, 2013) stress patterns to characterize features of the adaptive response to stress, such as the bandwidth or the accuracy of the response. Recently, we have used time-varying stress patterns to characterize both the response time of the hydrogen peroxide (H_2O_2) defense machinery (using linear ramps) and the emergence of acquired stress resistance (using a sequence of step-like stresses) (Goulev et al., 2017).

In bulk experiments, the generation of complex stress patterns can lead to potential artifacts. Media exchanges require multiple washing steps that may generate additional stress to the cells (e.g., temperature variations, desiccation, mechanical stress during centrifugation, etc.). In addition, the concentration of the stressor in the medium may be unstable due to active degradation by growing cells (e.g., H_2O_2 is highly unstable in a cell culture), see Fig. 1A. Last, exponential cell growth leads to nutrients depletion in the media, which limits the maximum duration of the experiments to a small number of cell divisions.

To overcome these issues, microfluidic devices have been increasingly used during the last decade to monitor the growth of individual cells exposed to a controllable environment (Bennett & Hasty, 2009; Charvin, Cross, & Siggia, 2008; Cookson et al., 2005; Velve-Casquillas et al., 2010). Indeed, one important advantage of microfluidics is to ensure a constant replenishment of the medium, thus preserving the stability of cellular environment over time. In addition, the possibility to switch media using computer-controlled electrovalves provides exquisite versatility that makes it particularly suitable to expose the cells to complex time-varying stimuli. Different stress patterns can be employed to investigate the cellular stress response (Fig. 1B). A simple switch from control to stress-supplemented media allows one to investigate the limits of the cellular adaptation to abrupt environmental changes. Alternatively, more complex sequences of medium switches can be generated (periodic pulses, staircases, etc.) see Fig. 1B. In addition, ramping stress can be

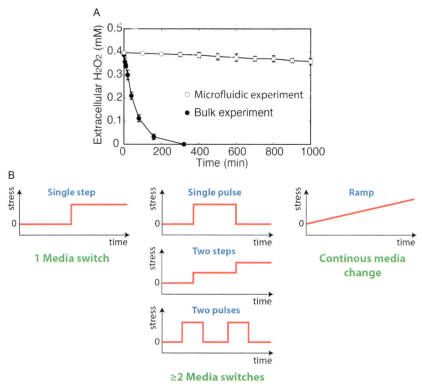

FIG. 1

(A) Stability of H_2O_2 in bulk vs microfluidics experiments. H_2O_2 concentration is measured using a colorimetric assay kit, as indicated in the text. (B) Example of possible media switches used in microfluidic devices.

Adapted from Goulev, Y., Morlot, S., Matifas, A., Huang, B., Molin, M., Toledano, M.B., et al. (2017). Nonlinear feedback drives homeostatic plasticity in H2O2 stress response. eLife, 6, e23971.

instrumental to reveal the magnitude of the stress homeostatic response (Goulev et al., 2017), to decipher the underlying architecture of the stress signaling network (Muzzey & van Oudenaarden, 2009) and to explore the possibility of stress rate-dependent responses (Young et al., 2013). Since microfluidic devices are fully compatible with imaging techniques, cellular response can be monitored with single cell resolution enabling the possibility to understand the origin of phenotypic cell-to-cell variability in stress response.

In this paper, we describe the methodology that we have developed to generate H_2O_2 stress patterns in a microfluidic device in order to study the adaptation of budding yeast cells to oxidative stress (Goulev et al., 2017). The microfluidic device is designed to ensure the accuracy of the stress pattern and to allow the simultaneous study of several independent stress patterns and mutant strains in parallel.

2 METHOD

2.1 INTRINSIC INSTABILITY OF H_2O_2

In bulk experiments, H_2O_2 is rapidly degraded by exponentially growing yeast cells (Fig. 1A). This makes it impossible to maintain a stable stress level over time. In contrast, in a microfluidic flow chamber, the constant replenishment of media considerably reduces the effect of cellular H_2O_2 consumption (Fig. 1A). Still, a small decay in H_2O_2 concentration is observed, yet <20% over a 1000-min period. This decay originates from the intrinsic instability of H_2O_2 in the tank that contains the medium. This decay depends on the type of media used for the experiment and is significantly slower for synthetic (SC) than for rich (YPD) cell media (not shown). The decay of H_2O_2 can be reduced to <10% over a 1000-min period by keeping the stock cell media on ice over the course of the experiment (Fig. 1A). In this case, one must ensure that the temperature of the medium is high enough when it reaches the cells in the chip. In our conditions (25 µL/min flow rate and 1 mm tube section), 30 cm of tubing are sufficient to thermalize the medium before it reaches the heated microfluidic chip.

3 MICROFLUIDIC CHIP DESIGN

The microfluidic chip, which is made of polydimethylsiloxane (PDMS), is designed to allow parallel independent experiments (with different strains exposed to separate environmental conditions), as displayed on Fig. 2A. To this end, the chip includes eight independent chambers, which are composed of two parallel media flow channels (50 µm high) that are connected to each other by eight micro-channels (length: 300 µm, width: 150 µm). Cell trapping in the micro-channels is ensured by their shallow depth (3.5 µm) which imposes a mechanical constrain on the cells. Therefore, yeast cells grow as a bi-dimensional monolayer and exceeding cells are washed away through the main flow channels.

In order to load cells into the micro-channels, two additional loading channels are connected to each chamber (Fig. 2A). These channels are parallel to the micro-channels thus facilitating the loading of cells. After successful cell loading, these two channels are clamped and remain so during the course of the experiment.

Nutrients and/or stressors reach the cell micro-colonies by diffusion. Using a fluorescent dye, we have checked that media diffusion through the micro-channels was not impaired by the presence of large cell clusters (compare Fig. 2B and C).

4 PDMS CHIP FABRICATION

Needed material:

– Microfluidics master
– PDSM (Sylgard 184)

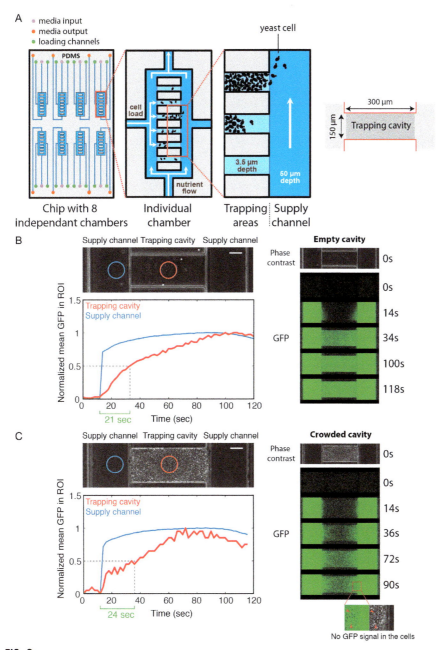

FIG. 2

(A) Sketch of the microfluidic device. (B) and (C) Diffusion of a FITC fluorescence dye through the micro-chambers used to trap cells in the absence (B) or presence of a dense cell cluster.

The image is adapted from Goulev, Y., Morlot, S., Matifas, A., Huang, B., Molin, M., Toledano, M.B., et al. (2017). Nonlinear feedback drives homeostatic plasticity in H2O2 stress response. eLife, 6, e23971.

- Curing agent (Sylgard 184)
- Desiccator
- Oven
- Biopsy puncher
- Coverslips
- Regular tape
- Plasma cleaner (Diener)

The microfluidics master is made using standard SU8 lithography on a silicon wafer (Whitesides, Ostuni, Takayama, Jiang, & Ingber, 2001). The lithography mask file used to generate the prototype is available upon request. The prototypic mold is replicated using epoxy glue to ensure long-term usage of the design (Xia et al., 1997).

The preparation of the PDMS chip consists of the following steps:

- Vigorously mix the PDMS with the curing agent (10:1 ratio). The polymerization efficiency of the final PDMS chip could be altered if the mixing is not efficient.
- Pour the PDMS on top of the mold to reach an ideal 5 mm thickness, then place it in a desiccator and start the vacuum pump. Wait until you see no more bubbles (roughly 30 min).
- Cure the PDMS at 80 °C for 4 h (master mold). If, instead of the silicone master, an epoxy replicate is used to cast the PDMS, the temperature should be lowered to 65 °C to prevent the softening of the epoxy replicate.
- Carefully cut the PDMS chip using a knife and peel it off. Make sure not to scratch the mold surface when cutting the PDMS.
- Punch holes in the PDMS chip using a biopsy puncher of appropriate size to make inlet and outlet connections (the puncher diameter should be slightly smaller than the diameter of tubing to prevent leaks).
- Wash and dry a coverslip using pressured air. Remove dusts and small chunks of PDMS from the surface of the chip using regular adhesive tape (this step is critical to ensure an efficient bonding to a glass coverslip, see below). Put both the coverslip and the PDMS chip (channel motifs face up) in a plasma cleaner to activate their respective surfaces. The plasma treatment parameters (time duration, power and pressure) must be appropriately set up to optimize bonding between PDMS and glass. (If the activation is too light, covalent bonding will not be homogenous at the PDMS/glass interface, see below. If too strong, the surface of the chip will be altered.)
- Put the PDMS chip on the coverslip with no delay after the plasma treatment. This step will covalently bind the PDMS chip to the coverslip, thus preventing any leakage when flowing media through the device channels.
- In order to improve the binding between the PDMS chip and the coverslip additional heating at 65 °C for 40 min is recommended. Do not exceed 40 min, otherwise the PDMS may become hydrophobic, which is detrimental to proper media flowing in the device.
- Immediately proceed with tubing connection and cell media flow through the chip, again, to prevent the PDMS from becoming hydrophobic.

5 MICROFLUIDICS LIVE-CELL IMAGING SETUP

The main components of the microfluidic setup are (Fig. 3):

– PDMS chip
– Epifluorescence microscope with focus control system (Zeiss Axio Observer Z1 + Definite Focus, Germany)
– Sample temperature regulation
– Tubing (Tygon or PTFE)
– Peristaltic pump
– Electrovalves (multiple electrovalves may be required to control media in independent chambers)
– Stock media (2% dextrose-complemented synthetic media)
– Colorimetric H_2O_2 assay kit (OxiSelect Hydrogen Peroxide Assay Kit (Colorimetric), STA-343, EUROMEDEX)

Time-lapse acquisition is achieved using an inverted epifluorescence microscope. To control the temperature of the cell environment (30 °C for yeast) a sample holder with thermoelectric modules is required as well as an objective heater (especially in the case of an immersion objective, which increases heat loss due to contact with the sample).

To drive media flow through the chip, a peristaltic pump, a syringe pump or a pressure controller can be used. A flow rate of typically 5–50 µL/min is sufficient to ensure constant supply of nutrients to the cells. Higher flow rates increase the risk of leaks from the channels due to increasing pressure. Media switches are automated

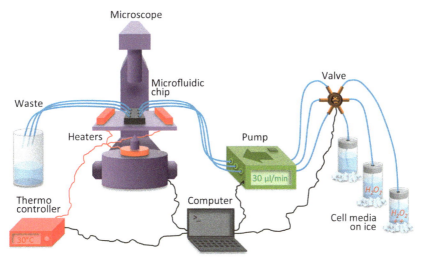

FIG. 3

Overview of the setup used to performed controllable stress patterns and simultaneous image acquisition on an inverted epifluorescence microscope.

using a computer-controlled electrovalve, which enables the accurate generation of complex stress patterns.

Regarding the choice of media, hydrogen peroxide is considerably more stable in synthetic complete (SC) media than in rich YPD media. In addition, YPD is highly auto-fluorescent, which makes it unsuitable for fluorescence microscopy. Therefore a 2% dextrose-complemented SC media (SCD) is ideal for microfluidics experiments involving H_2O_2 stress. Maintaining media stock on ice during the course of the experiments limits the degradation of H_2O_2, as discussed above.

Protocol to set up a time-lapse experiment:

– Grow yeast overnight in SCD medium (at 30°C, 220 RPM).
– On the next morning, dilute the overnight culture to 0.1–0.2 optical density (OD) and incubate for another 4 h to reach a final 0.4–0.6 OD. A lower density will affect the efficiency of cell loading. Conversely, the cellular fitness may be affected if the cellular density is too high.
– Prepare about 1 L of SCD media (without H_2O_2) and 1 L of SCD media containing H_2O_2 at the desired concentration and keep them on ice.
– Connect the different media tanks to the inlets of electrovalve(s) using appropriate tubing.
– Connect the outlet of the electrovalve(s) to a peristaltic pump and then to the individual chambers of the PDMS microfluidic chip.
– Connect the outlet of the PDMS microfluidic chip to a waste tank.
– Start the pump and load the chambers with SCD without H_2O_2.
– Use the side channels (loading channels) of the chambers to manually load the cells into the PDMS microfluidic chip with a syringe/needle (during this step the pump should be turned off).
– Clamp the loading channels and restart the pump.
– Start the time-lapse acquisition.
– Use the electrovalve(s) to switch media and to generate appropriate H_2O_2 stress patterns.
– To ensure the accuracy of the H_2O_2 stress exposure, samples of waste media can be periodically taken from the outlet of the microfluidic chip to check the H_2O_2 concentration during the course of the experiment. The H_2O_2 concentration can be measured using a colorimetric H_2O_2 assay kit.

6 PRINCIPLES OF H_2O_2 RAMPING STRESS

In addition to step-like stress exposures, stress ramps provide a useful mean to assess the homeostatic properties of a stress response pathway. Linear H_2O_2 stress ramp can be simply generated by constantly pumping concentrated H_2O_2 from a stock solution to the tank that contains the cell growth medium (Fig. 4A). In this case, a magnetic stirrer ensures proper medium homogenization.

Under certain assumptions the temporal profile of H_2O_2 concentration obtained with this method is close to linear. Let C_1 and V_1 be the concentration and volume of

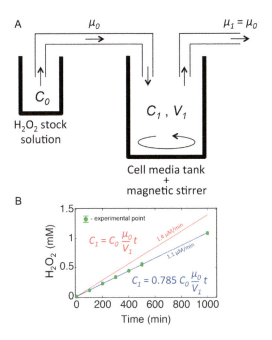

FIG. 4

(A) Generic principle of a media mixing strategy used to generate a temporally controlled concentration of a chemical compound (e.g., H$_2$O$_2$). (B) Kinetic of H$_2$O$_2$ concentration in the microfluidic device obtained using the method described in (A).

Adapted from Goulev, Y., Morlot, S., Matifas, A., Huang, B., Molin, M., Toledano, M.B., et al. (2017). Nonlinear feedback drives homeostatic plasticity in H2O2 stress response. eLife, 6, e23971.

the media tank, respectively (Fig. 4A). If μ_1 is the flow rate used to perfuse the microfluidic device from the tank, and if C_0 and μ_0 are the concentration and flow rate from the H$_2$O$_2$ stock solution, respectively, the evolution of V_1 and C_1 can be written as:

$$\frac{dV_1}{dt} = \mu_0 - \mu_1$$

$$\frac{dC_1}{dt} = \frac{\mu_0}{V_1^0 + (\mu_0 - \mu_1)t}(C_0 - C_1)$$

When $\mu_0 = \mu_1$ (identical flow rates), V_1 is constant and the evolution of C_1 with time is given by:

$$C_1 = C_0\left(1 - e^{\delta t}\right)$$

where $\delta = \frac{\mu_0}{V_1}$. When $t \ll 1/\delta$, then the concentration C_1 increases linearly as:

$$C_1 = C_0 \delta t$$

Assuming $\mu_0 = 30\,\mu\text{L/min}$ and $V_1 = 1\,\text{L}$, we obtain $\delta = 3 \times 10^{-5}\,\text{min}^{-1}$. In this case, using $C_0 = 46.67\,\text{mM}$, one expects a linear slope of $1.4\,\mu\text{M/min}$, provided that

$t \ll 10^5$ min (Fig. 4B, see red curve). Experimental measurements revealed that the evolution of concentration with time was indeed linear (Fig. 4B). However, H_2O_2 concentration dosages showed that, while the ramping was quite linear, the actual concentration was $21.5\% \pm 2\%$ lower than expected over a range of stress rates 1.4–$22.4\,\mu M/min$ (Goulev et al., 2017). Therefore, the expression for C_1 has to be corrected by a factor $a = 0.785$ such that (Fig. 4C, see blue curve):

$$C_1 = aC_0\delta t$$

The deviation to the expected concentration is likely to originate from the spontaneous degradation of H_2O_2 in the mixing tank, which undergoes constant stirring and hence cannot be kept on ice. Therefore, depending on the particular conditions of the ramp experiments, including media chemical composition, room temperature etc., it is highly recommended to calibrate the value of a before setting up a time-lapse assay.

7 STRESS RAMPING SETUP

In order to perform H_2O_2 ramping, the following components must be added to the experimental setup (Fig. 4B):

– An additional media tank (a large volume ($>1\,L$) is needed to ensure a linear concentration increase over a long timescale, see above)
– A H_2O_2 stock solution (highly concentrated, i.e., 30–$800\,mM$, in SCD media)
– A magnetic stirrer
– A second peristaltic pump (or alternatively, a multi-channel pump). In case a second pump is used, it may be necessary to calibrate the flow rate to ensure that $\mu_0 = \mu_1$).

The protocol to perform ramping H_2O_2 stress is as follows:

– Calculate the volume of the media tank and the concentration of the H_2O_2 stock solution according to the desired slope of the H_2O_2 ramp as indicated above. (The flow rate of the H_2O_2 stock solution entering the media tank should be the same as the flow rate of media coming out of the media.)
– Prepare the media tank with the appropriate volume of SCD without H_2O_2.
– Prepare H_2O_2 stock solution.
– Put the media tank on a magnetic stirrer and start the mixing.
– Connect the H_2O_2 stock solution to the media tank.
– Connect the media tank to the microfluidic PDMS chip.
– After loading the cells, start the pump and the time-lapse acquisition.

In case the ramp should not start right at the beginning of the time-lapse acquisition, the H_2O_2 solution that enters the media tank has to be initially replaced by plain SCD medium. To start the ramp, use an electrovalve to select the SCD media with the H_2O_2 stock solution.

8 DISCUSSION

In this paper, we provide a detailed protocol used to generate controllable temporal stress patterns in a microfluidic device designed for multi-generation cell growth assays. Using this methodology, the adaptive properties of yeast cells can be investigated with good temporal resolution on physiological timescales. Specifically, a 3-min imaging interval allows one to compute the instantaneous changes in growth rate occurring in response to a pulse of stressor, which is a key marker of physiological adaptation (Goulev et al., 2017; Young et al., 2013). In addition, single cell resolution provides an essential mean to assess the variability in cellular response, which is required to understand complex adaptation schemes (Levy, Ziv, & Siegal, 2012; Veening, Smits, & Kuipers, 2008) at the population level in unicellular species.

The goal of our previous study was to elucidate functional properties associated with the adaptation to hydrogen peroxide (Goulev et al., 2017), hence this protocol exclusively focuses on this particular stressor. However, most of the procedures described here may apply to other types of stress, including carbon source switch, starvation, pH or any drug that alters a specific biological process.

There are two essential points to ensure the accuracy and the reliability of these experiments: first, it is necessary to assess whether the stability of the drug/stressor is compatible with multi-generation timescale imaging. In the case of hydrogen peroxide, the degradation of this compound can be limited by keeping the medium on ice, and it is likely that this precaution applies to other cases. Second, it has been reported that PDMS may adsorb chemical compounds and thus alter the effect of a given stimulus (Chen et al., 2016). Therefore, it is mandatory to check that experiments done in PDMS chips lead to comparable results as those performed in test tubes. If it is not the case, alternative materials have been proposed to design micron-scale channels suitable for cell growth assays (Chen et al., 2016).

The ability to perform a number of independent assays with different stress patterns or strains considerably increases the throughput of the experiment. However, there is a trade-off between the number of fields of view imaged during a time-lapse sequence and the duration of the time-lapse interval. Single cell analyses impose a time-lapse interval lower than typically 10 min, otherwise individual cell tracking cannot be performed accurately (Fehrmann et al., 2013). Custom hardware modifications (Dénervaud et al., 2013) of the image acquisition pipeline and improvements in the image processing workflows should further increase the accuracy and throughput of cell growth quantification in microfluidic devices.

REFERENCES

Auesukaree, C. (2017). Molecular mechanisms of the yeast adaptive response and tolerance to stresses encountered during ethanol fermentation. *Journal of Bioscience and Bioengineering*, *124*(2), 133–142.

Bennett, M. R., & Hasty, J. (2009). Microfluidic devices for measuring gene network dynamics in single cells. *Nature Reviews Genetics*, *10*(9), 628–638.

Castillo-Hair, S. M., Igoshin, O. A., & Tabor, J. J. (2015). How to train your microbe: methods for dynamically characterizing gene networks. *Current Opinion in Microbiology, 24,* 113–123.

Charvin, G., Cross, F. R., & Siggia, E. D. (2008). A microfluidic device for temporally controlled gene expression and long-term fluorescent imaging in unperturbed dividing yeast cells. *PLOS One, 3*(1), e1468.

Chen, T., Gomez-Escoda, B., Munoz-Garcia, J., Babic, J., Griscom, L., Wu, P.-Y. J., et al. (2016). A drug-compatible and temperature-controlled microfluidic device for live-cell imaging. *Open Biology, 6,* 160156.

Cookson, S., et al. (2005). Monitoring dynamics of single-cell gene expression over multiple cell cycles. *Molecular Systems Biology, 1,* 2005.0024.

Davies, K. J. (2016). Adaptive homeostasis. *Molecular Aspects of Medicine, 49,* 1–7.

Dénervaud, N., Becker, J., Delgado-Gonzalo, R., Damay, P., Rajkumar, A. S., Unser, M., et al. (2013). A chemostat array enables the spatio-temporal analysis of the yeast proteome. *Proceedings of the National Academy of Sciences of the United States of America, 110,* 15842–15847.

Fehrmann, S., Paoletti, C., Goulev, Y., Ungureanu, A., Aguilaniu, H., & Charvin, G. (2013). Aging yeast cells undergo a sharp entry into senescence unrelated to the loss of mitochondrial membrane potential. *Cell Reports, 5,* 1589–1599.

Goulev, Y., Morlot, S., Matifas, A., Huang, B., Molin, M., Toledano, M. B., et al. (2017). Nonlinear feedback drives homeostatic plasticity in H_2O_2 stress response. *eLife, 6,* e23971.

Guan, Q., Haroon, S., Bravo, D. G., Will, J. L., & Gasch, A. P. (2012). Cellular memory of acquired stress resistance in Saccharomyces cerevisiae. *Genetics, 192,* 495–505.

Hersen, P., et al. (2008). Signal processing by the HOG MAP kinase pathway. *Proceedings of the National Academy of Sciences of the United States of America, 105*(20), 7165–7170.

Hohmann, S. (2002). Osmotic adaptation in yeast-control of the yeast osmolyte system. *International Review of Cytology, 215,* 149–187.

Levy, S. F., Ziv, N., & Siegal, M. L. (2012). Bet hedging in yeast by heterogeneous, age-correlated expression of a stress protectant. *PLoS Biology, 10,* e1001325.

Mettetal, J. T., et al. (2008). The frequency dependence of osmo-adaptation in Saccharomyces cerevisiae. *Science, 319*(5862), 482–484.

Morano, K. A., Grant, C. M., & Moye-Rowley, W. S. (2012). The response to heat shock and oxidative stress in Saccharomyces cerevisiae. *Genetics, 190*(4), 1157–1195.

Muzzey, D., & van Oudenaarden, A. (2009). Quantitative time-lapse fluorescence microscopy in single cells. *Annual Review of Cell and Developmental Biology, 25,* 301–327.

Veening, J.-W., Smits, W. K., & Kuipers, O. P. (2008). Bistability, epigenetics, and bet-hedging in bacteria. *Annual Review of Microbiology, 62,* 193–210.

Velve-Casquillas, G., et al. (2010). Microfluidic tools for cell biological research. *Nano Today, 5*(1), 28–47.

Whitesides, G. M., Ostuni, E., Takayama, S., Jiang, X., & Ingber, D. E. (2001). Soft lithography in biology and biochemistry. *Annual Review of Biomedical Engineering, 3,* 335–373.

Xia, Y., McClelland, J. J., Gupta, R., Qin, D., Zhao, X.-M., Sohn, L. L., et al. (1997). Replica molding using polymeric materials: A practical step toward nanomanufacturing. *Advanced Materials, 9,* 147–149.

Young, J. W., Locke, J. C. W., & Elowitz, M. B. (2013). Rate of environmental change determines stress response specificity. *Proceedings of the National Academy of Sciences of the United States of America, 110,* 4140–4145.

A bacterial antibiotic resistance accelerator and applications

3

Julia Bos*, Robert H. Austin[†],[1]

*Pasteur Institute, Department of Genomes and Genetics, Paris, France
[†]Department of Physics, Princeton University, Princeton, NJ, United States
[1]Corresponding author: e-mail address: austin@Princeton.EDU

CHAPTER OUTLINE

Abstract

The systematic emergence of drug resistance remains a major problem in the treatment of infectious diseases (antibiotics) and cancer (chemotherapy), with possible common fundamental origins linking bacterial antibiotic resistance and emergence of chemotherapy resistance. The common link may be evolution in a complex fitness landscape with connected small population niches. We report a detailed method for observing bacterial adaptive behavior in heterogeneous microfluidic environment designed to mimic the environmental heterogeneity found in natural microbial niches. First, the device is structured with multiple connected

Methods in Cell Biology, Volume 147, ISSN 0091-679X, https://doi.org/10.1016/bs.mcb.2018.06.005

micro-chambers that allow the cell population to communicate and organize into smaller populations. Second, bacteria evolve within an antibiotic gradient generated throughout the micro-chambers that creates a wide range of fitness landscapes. High-resolution images of the adaptive response to the antibiotic stress are captured by epifluorescence microscopy at various levels of the bacterial organization for quantitative analysis. Thus, the experimental setup we have developed provides a powerful frame for visualizing evolution at work: bacterial movement, survival and death. It also presents a basis for exploring the rates at which drug resistance arises in bacteria and other biological contexts such as cancer.

1 INTRODUCTION

The emergence of bacterial antibiotic resistance is a growing problem, responsible for over 700,000 deaths worldwide (Jasovsky, Littmann, Zorzet, & Cars, 2016). The development of resistance is not going to go away as long as the overuse and misuse of antibiotics are common practice, and even if we do rein in present abuses the treat of the emergence of resistance will always be present. There is an urgent need of gaining knowledge on the mechanisms underlying the origin of resistance to be able to slow down the evolution of resistance.

Bacteria develop in complex ecologies: they arrange themselves into emergent structured, mixed, sessile communities (i.e., biofilms (Flemming et al., 2016; Lebeaux, Ghigo, & Beloin, 2014)) and evolve in compartmented ecologies where they respond to a variety of external gradients, such as, e.g., nutrient, chemical, oxygen, or temperature gradients (Adler, 1975; Berg, 1975; Borer, Tecon, & Or, 2018; Fenchel & Finlay, 2008; Hu & Tu, 2013, Jiang, Ouyang, & Tu, 2009; Khan, Spudich, McCray, & Trentham, 1995; Maeda, Imae, Shioi, & Oosawa, 1976; Paster & Ryu, 2008; Salman & Libchaber, 2007; Taylor, Zhulin, & Johnson, 1999; Yang & Sourjik, 2012; Yu, 2002). So far, our ability to understand the physiology and evolution of bacteria in their natural habitat is far from satisfactory. Most fundamental research in microbiology over the past half-century has mostly ignored these organizational properties of bacterial communities by using homogeneous and simply structured environments, such as agar plates, agitated test tubes or flasks as growth mode. Although the use of these traditional methods has largely contributed to massive identification and characterization of the molecular pathways as well as the genetic changes in the bacterial genomes that mediate low level stress response and adaptation (Good, McDonald, Barrick, Lenski, & Desai, 2017), it has poorly contributed to comprehend the origins of antibiotic resistance under high-stress and complex conditions, i.e., in the real world.

However, in the last decade, the development of micro-fabricated structures and microfluidic technology combined with microscopy has offered the possibility of the study of many long-standing questions concerning the organizational and evolutionary aspects of microbial communities (Adler & Groisman, 2012; Aizel et al., 2017; Amselem, Guermonprez, Drogue, Michelin, & Baroud, 2016; Coyte, Tabuteau, Gaffney, Foster, & Durham, 2016; Crooks, Stilwell, Oliver, Zhong, & Weibel, 2015; Hol & Dekker, 2014; Morris et al., 2017; Weibel, Diluzio, & Whitesides, 2007; Wessel, Hmelo, Parsek, & Whiteley, 2013; Zhang et al., 2014).

To address the need for a better comprehension of bacterial strategies for adaptation against antibiotics under more realistic high-stress/complex ecology environments, we developed a microfluidic cell culture system that allows recording time lapse images of the bacteria cell behavior in a drug gradient and a complex metapopulation (Keymer, Galajda, Muldoon, Park, & Austin, 2006; Zhang, Lambert, et al., 2011; Zhang, Robin, Liao, Lambert, & Austin, 2011). The bacterial population in these devices is actually a metapopulation subdivided into multiple interconnected micro-habitats. Such spatial geometry of the array enables semi-restricted movement between hexagons, microhabitat–microhabitat communication and population fragmentation into smaller sub-populations. This recapitulates Darwinian natural selection in natural environments with possible changes in selective pressures over time and space. Nutrients and antibiotics are delivered to the microhabitats via short nanoslits at the edge of the periphery channels which minimize advective flow into the ecology. Bacteria developing in the array are thus exposed to a gradient of antibiotic that generates spatial opportunities for genetic and/or non-genetic changes that increase fitness. Therefore, our device provides a way to quantitatively visualize the behavior of bacterial populations evolving in structured microenvironments in response to an antibiotic gradient.

In the protocol described in this review, we specifically applied a gradient of ciprofloxacin antibiotic, a commonly used antibiotic that generates DNA damage and makes bacteria filamentous by blocking cellular division (Davis, Markham, & Balfour, 1996). Bacterial growth was monitored by measuring the fluorescence intensity of the local bacterial population over time at various spatial resolutions, enabling both overall global populations and individual bacterial imaging.

Such experimental setup enabled us demonstrate that (i) bacteria can evolve ciprofloxacin resistance at faster rates in structured microenvironments than if cultured in traditional flasks or test tubes (Zhang, Lambert, et al., 2011; Zhang, Robin, et al., 2011) and (ii) rapid adaptation occurs via transient filamentation stage that gives birth to the final highly resistant bacteria in heterogeneous landscapes (Bos et al., 2015).

From this initial set of experiments, we anticipate that the dissection of the behavioral and mechanistic steps toward the development of resistance in custom fabricated devices will provide a strong technological platform for the comprehension of how bacteria and more complex cellular systems like cancer cells, are capable of altering themselves genetically to rapidly increase their fitness in a high-stress environment.

2 MATERIALS

2.1 MICROFLUIDICS

- 100 mm silicon wafers
- AZ 5214 and AZ MIF300 Photoresists
- MA6 mask aligner

- Samco 800 (etching)
- TePla M4L plasma etcher (resist strip)
- PDMS Silgard 184 (Dow Corning Midland, MI)
- 24 mm × 60 mm glass coverslips
- Soft tubing (Tygon 0.06″ ID)
- Hollow stainless steel pins (0.036″ OD, 0.023″ ID, Small Parts Inc.)
- Silicone O-rings (McMaster-Carr Robbinsville, NJ 08691)
- Acrylic sample holder
- Syringe pump (Chemyx or Harvard apparatus)
- Furnace for thermal oxidation
- Plasma Preen (North Brunswick, NJ) plasma equipment to bond PDMS to glass

2.2 GROWTH MEDIUM AND REAGENTS

- LB broth (Miller; Sigma) stored at room temperature
- Ciprofloxacin (Sigma)
- Fluorescein (Sigma)
- Sterile culture tubes (Eppendorf)
- Gel-loading pipet tips
- Acetone
- Ethanol
- Deionized water

3 METHODS

3.1 DEVICE FABRICATION AND CLEANING

The overview of our original chip array design is presented in Fig. 1. The structure (approx. 2 cm × 3 cm) is etched into a 100 mm silicon wafer and sealed on the top with a PDMS-layered glass coverslip. The micro-ecology consisted of 1200 chambers shaped as hexagonal habitats, each habitat had six sides 200 μm long and was etched 10 μm deep into the silicon (Fig. 1). The chambers allow bacterial movement between adjacent microhabitats via long and narrow microchannels that are 200 μm long × 10 μm wide × 10 μm deep (Fig. 1). The hexagonal shape of the chambers was picked to optimize both the number of confined habitats along with the number of connectors in between the chambers. Wide periphery channels surrounding the microhabitats diffuse in food and antibiotics through the microhabitat array via 100 nm deep nanoslits etched into the sidewalls of the feeding channels.

The microfluidic device was fabricated by photolithography and reactive dry etching. The two-layer device consisted of a deep-etched (10 μm) layer that contains the hexagonal array and microchannels for culturing cells and delivering food, and a shallow layer (100 nm) that contains the nanoslits which isolate the culture chambers

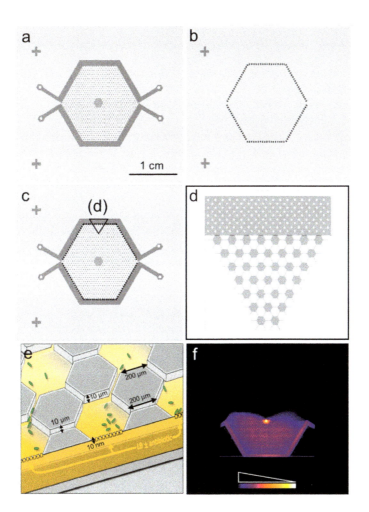

FIG. 1

Design of our microfluidic array (A). The chip is composed of 1200 hexagonal wells etched 10 µm deep into a silicon wafer. The nanoslits pattern (B) is aligned on the microhabitat array wafer (A) and exposed to UV light to produce the final pattern in (C) using standard photolithography. Details of the microhabitat pattern are shown up in (D) and (E). Nanoslits 100 nm deep are etched in the sidewalls of the feeding channel at the edge of the array allowing nutrients and/or antibiotic to flow into the device. Each side of a microchamber measures 200 µm long and is connected to its nearest neighbors via six microchannels that are 200 µm long, 10 µm deep, and 10 µm wide. Growth medium containing or not the antibiotic flows in the periphery channels and enters the array through the nanoslits. The generation of a complex gradient within the array is shown in (F) with 0.2 mM fluorescein dye running in one side of the chip only.

Panel (F): Modified from Zhang, Q., Lambert, G., Liao, D., Kim, H., Robin, K., Tung, C. K., et al. (2011). Acceleration of emergence of bacterial antibiotic resistance in connected microenvironments. Science, *333(6050), 1764–1767; Zhang, Q., Robin, K., Liao, D., Lambert, G., & Austin, R. H. (2011). The goldilocks principle and antibiotic resistance in bacteria.* Molecular Pharmaceutics, *8(6), 2063–2068. https://doi.org/10. 1021/mp200274r (web archive link).*

from the fluidic feeding channels. Perfect alignment of the two layers is critical for proper fabrication process. The fabrication and cleaning steps are the following:

1. The 4″ silicon wafer was spin coated with photoresist AZ 5214, resulting in a uniform thin layer.
2. The photo resist-coated wafer was baked at 100 °C for 30–60 s on a hotplate to remove excess photoresist solvent.
3. The photo resist-coated wafer was aligned with photomask containing the nanoslits and then exposed to UV light in a Karl Suss MA6 mask aligner.
4. The exposed photoresist was developed in AZ MIF300.
5. Etching process: the nanoslits were etched 100 nm deep using a Samco 800 machine.
6. The photoresist was removed from the substrate by soaking the wafer in acetone for 10 min. These steps were repeated with the microhabitats mask, instead the wafer was etched 10 μm deep and a protective layer of photoresist was added to the surface after the last etching step before laser dicing of peripheral and central ports.
7. Five holes (1 mm wide) were drilled in the inlets and outlets ports, as well as in the center port for bacterial inoculation, using laser dicing.
8. Solvent clean: the chip was soaked in acetone (10 min) to remove the photoresist layer. A final strip of residual photoresist was removed in a TePla M4L plasma etcher with an oxygen plasma. Careful cleaning of the chip is crucial for reproducible results.
9. Oxide growth: since PDMS covalently bonds to bare silicon, preventing re-use of the chips because of adhesion of the PDMS-coated coverslip, a 0.5 μm layer of silicon oxide was grown at the surface substrate by exposure of the wafer for 2 h at 1000 °C in an oxygen atmosphere.
10. Lid preparation: PDMS mixture was degassed in a vacuum chamber for 30 min before being spun onto the coverslip 24 mm × 60 mm glass coverslip (4000 rpm for 10 min) and heat cross linked at 60 °C for 1 h to obtain a thin layer to seal the silicon chip array. All steps were performed under a laminar flow hood to avoid any particles sticking to the PDMS surface.
11. Chip wetting: before sealing, the PDMS-coated coverslip was exposed to 20 s plasma treatment (Plasma Preen) to increase wetting by creating a transient hydrophilic surface. Immediately after plasma treatment the coverslip was pressed against the silicon chip. Next, the chip was wet with LB medium by vacuum immersion at room temperature. The device was immersed in a sterile glass container filled with LB medium and vacuum (−1 bar) was applied for 30 min. Slow pressure (10 min from vacuum to atmospheric pressure) release was crucial for proper wetting.
12. Syringes and tubing were next filled with appropriate medium (with or without the cipro antibiotic) and connected to the device directly in the device via pre-purged Tygon tubes to avoid introducing air bubbles into the array. The device and syringes and tubing was transported as is to the inverted microscope and syringe pump.

3.2 **ASSEMBLY OF THE DEVICE**

Assembly of the microfluidic device is illustrated in Fig. 2. Our experimental platform can be installed on any inverted or upright microscope as it requires only a peristaltic pump and the portable microfluidic device.

1. Nutrients and antibiotic delivery to the device: a syringe pump is programmed to both deliver growth medium supplemented or not with the antibiotic and withdraw waste in a continuous way. Four 3 mL BD Luer Lock™ syringes are loaded onto the syringe pump and individually connected to hollow steel pins inserted in the sample holder through soft Tygon™ tubing. Each steel pin connects to a port on a chip. O-rings ensure proper sealing of the contact between ports and steel pins (Fig. 2). Two syringes filled with sterile LB medium containing different concentrations of cipro antibiotic (typically 0 and 1 µg/mL) were loaded in the infusion deck of the syringe pump, whereas two other empty

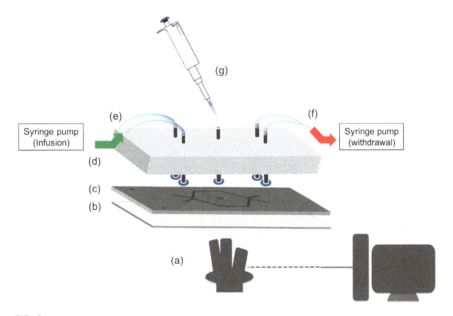

FIG. 2

Microfluidic chip assembly. The silicon chip array (C) is covered with a PDMS-layered glass coverslip (B) prior being mounted onto a plastic-based chip holder (D) in which are inserted the ports-connected hollow steel pins. O-rings (dark blue) are used as joints to ensure effective sealing (air- and fluid-free) at the interface between the silicon wafer and the tip of the hollow steel pins. Culture medium flows in (green arrows (E)) and out (red arrows (F)) of the array via soft tubing that connects the inlet/outlet ports on the chip to the medium supply (infusion syringes) and medium waste (withdrawal syringes), respectively. Fluorescent bacteria are loaded in the center chamber with the use of a pipetman and conical loading tips (G). The entire platform is set on a motorized stage of an inverted microscope (A).

syringes were placed in the withdrawal deck to harvest waste. While our protocol uses LB we have also experimented with minimal media to reduce autofluorescence of the medium and begin studies on the influence of the richness of the bacterial medium. The typical volumetric flow rate programmed into the syringe pump interface was set at 15 μL/h. In our experiments, a continuous flow of LB medium +/− ciprofloxacin antibiotic was achieved by the pressure difference between the loading reservoirs and the waste outlets.

2. Antibiotic gradient check: fluorescein dye (0.1 mM) was flowed in one side of the chip while the other side received LB only, to demonstrate that a stable gradient of ciprofloxacin is established after 24 h (Fig. 1F). Therefore, for each experiment we let the gradient of antibiotic run for 24 h prior to cell loading in the chip.

3.3 CELL CULTURE AND INOCULATION OF THE CENTRAL CHAMBER

1. Bacterial cell cultivation: fluorescent *Escherichia coli* cells were streaked out on a plate for isolation. After overnight growth, one colony was picked and cultured in plain LB medium (2 mL) (supplemented with IPTG (0.5 mM) to induce cell body fluorescence) for 2 h at 37°C until mid-exponential phase is reached (10^8 2cells/mL).

2. Inoculation of bacterial cell sample into the device: a small volume of bacteria cells (5 μL, approximately 10^5 cells) was placed in the open loading hole (Fig. 2). To prevent dryness entering the device via the loading hole, a pad of PDMS of about 1 mm thick was slid over the loading hole after placement of the bacterial sample.

3.4 REAL TIME IMAGING

1. Images of the whole array were taken using a Canon 5D camera using an MPE-65 macro lens. The excitation source was a 470 nm LED from Thorlabs. High-magnification images of individual micro-chambers were taken using a Canon 5D camera attached to a on a Nikon inverted TI-E microscope using a 40× CFI Plan Fluor (NA 0.75; WD 0.66).

2. Image acquisition: Micromanager software (https://micro-manager.org) was used to control the microscope and camera settings and to acquire time lapse images. Images of the chip array at various magnifications (macrolens and 40×) are shown in Fig. 3.

3.5 DOWNSTREAM ANALYSIS

1. Harvesting resistant bacterial population: typically, at the end of the experiment, the lid is gently removed from the array and the chip array is plated out on a Petri dish to isolate the antibiotic-resistant cells. After 24 h of growth at 37°C, cipro-resistant colonies are visible.

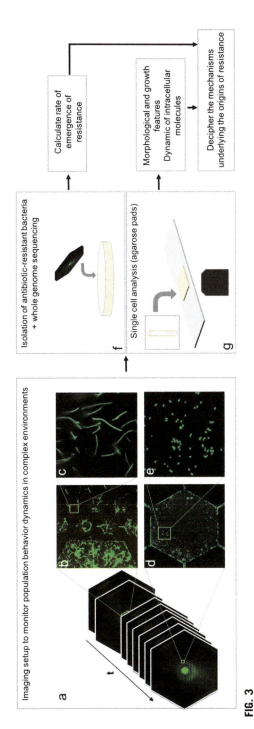

FIG. 3

Real time images of the bacterial population. Images were processed using micromanager software and imported into Image J software (http://imagej.nih.gov/ij/index.html). The whole chip was imaged with a macro lens objective (A). Image of bacteria thriving inside a microhabitat; evolved bacteria (B) or filamentous bacteria (C) were obtained with a macrolens and 40 × objective, respectively.

2. Isolating sub-population(s) of interest in a micro-chamber. We have put a fair amount of work at improving the device design and assembly with the goal of widening the opportunities for downstream analysis of the various fitness landscapes. We came up with an alternative assembly protocol and chip geometry that facilitated both sealing step and imaging process. We now use a type of lid and sealing technology such as a 35 mm diameter hydrophilic 20 μm thick Lumox™ gas-permeable membrane (Sarstedt AG & Co. KG), that renders the device capable of generating time-dependent atmosphere gradients (Lin et al., 2017). To ensure seal integrity, a gas supply system that pressurizes the space between the membrane and the device and pushes membrane against the array. In addition to offer air-tight sealing, the membrane lid presents a great potential for harvesting sub-populations of interest at a given time and position in the array as it is possible to poke through the membrane with a micro pipetting robot.

3. Biological analysis of the evolved population(s).

The cell samples can further be used for various biological analysis such as whole-genome sequencing (i.e., DNA and RNA seq) and metabolomics analysis in order to identify respectively, mutations, epigenetic changes and small compounds released in the medium that may influence the rate of the emergence of antibiotic resistance.

4 DEVICE PERFORMANCES

4.1 EMERGENCE OF ANTIBIOTIC RESISTANCE IN BACTERIAL POPULATIONS AND SINGLE CELLS

Emergence of cipro-resistant bacterial population occurred within less than 24 h of growth in the device. Following inoculation into the center of the device, the bacteria quickly consumed nutrients locally and gathered toward regions of high nutrient content at the perimeter of the device against the nanoslits. The first signs of development of resistant microorganisms arise along with bursts of bacterial growth at a local microenvironment which we call "Goldilocks points" (Zhang, Robin, et al., 2011), where the gradient between medium that contains and lacks the antibiotic is sharpest. Increased fitness of antibiotic-adapted cells is typically followed by propagating waves through the entire device of a mixture of antibiotic-sensitive and resistant bacteria competing for resources. The bacterial waves aim at colonizing unoccupied replenished micro-niches and led to complex growth dynamics pattern (Lambert, Liao, & Austin, 2010). When the waves stop propagating, the bacterial sub-populations eventually settle into the micro-chambers and grow biofilm-like structures. Deciphering the biological significance of the population(s) movement remains one of our current focus in the lab.

What we learned from these experiments is that both the drug gradient and the connected microenvironments play a key role in the rapid emergence of ciprofloxacin resistance. Connected microhabitats force the population to break up into sub-populations while maintaining motility between chambers. Such spatial geometry of the device along with the drug gradient imposed on the connected array allowed for quicker fixation of ciprofloxacin-adapted mutants in smaller populations and their move toward niches exposed to higher stress where fewer individuals are present. A control experiment where bacterial populations had been cultured in discrete chambers with varying concentrations of cipro showed no evolution of high resistance against ciprofloxacin antibiotic (Zhang, Lambert, et al., 2011; Zhang, Robin, et al., 2011). These findings provide more evidence that physical movement of bacterial populations across structured space is key to survival and spread in challenging environments.

Although the design of the experiment described earlier enabled quantitative monitoring of the population dynamics under a drug gradient, it had limitations in acquiring high spatial resolution time course images of bacterial cells. Therefore, to delineate the stages in the development of ciprofloxacin resistance, we performed high-resolution imaging where individual cells were visualized inside the microenvironments using a CFI Plan fluor (NA 0.75; WD 0.66) 40× lens. In such experimental setup, fluorescent bacteria were grown overnight in LB in test tubes and sub-cultured the next day in fresh LB. The culture was stopped when the cells had reached the exponential phase of growth and a small sample (2–5 μL of 10^6/mL) was used to inoculate the center of the chip once the gradient of cipro was established in the whole array. Images were taken every 10 min for several hours after inoculation.

High-resolution imaging of the microhabitats provided us with mechanistic details of development of resistance to low concentrations of antibiotic at the single-cell level. For instance, it allowed us to visualize in real time living bacteria exiting from the stressed state (i.e., filamentous) and resuming normal cell growth properties (Bos et al., 2015). It was also possible to push the analysis further by examining over shorter time frames the chromosome dynamics inside the evolution compartment, by immobilizing filamentous bacteria on solid cipro-containing agar pads. Filamentous bacteria generate multiple "mutant" chromosomes as they grow under antibiotic pressure, until a solution to overcome the antibiotic stress is obtained: normal-sized cells containing mutant chromosomes are pinched off of the filament tips and if resistant, give rise to normal-sized progeny which resume division (Bos et al., 2015).

Overall, this novel cultivation technique for bacteria in microhabitat patches permits to watch bacteria evolving real time resistance and thus paves the way for studying the complex behavior of bacterial populations in controlled environments that can mimic their natural habitats.

4.2 APPLICATIONS TO CANCER

As we have discussed, conventional cell culture was developed almost a century ago and remains the most frequently used preclinical model in biomedical research, despite its proven limited ability to predict clinical results in cancer. Conventional cell cultures are not able to recapitulate key components and interactions of the tumor microenvironment in a comprehensive manner while providing reliable and reproducible data. 3D cell cultures also have notable limitations, and their application for cancer drug screening remains anecdotal. Animal models remain the most effective preclinical platform to predict clinical response, but the goal of patient-derived xenografts to personalize therapies also remains anecdotal. Uncovering key interactions between host cells and cancer cells and developing improved therapeutic strategies under the conditions found in vivo requires not only a cell culturing system with complex fitness gradients in which the behaviors of each cell and the interactions of multiple cell types can be tracked and monitored in real time, but also a means to localize cell populations transiently so that local interactions can take precedence over "mean field" interactions averaged over all cell types.

For cancer work, the nanoslits used in the bacterial work are not necessary because of the size of cancer cells. We utilize standard photolithography and soft lithography technology to fabricate the microenvironment, which is composed of 109 interconnected hexagonal microhabitats, surrounded by two independent periphery medium channels around which we can circulate both nutrients and drugs. The microenvironment is sealed on the bottom of a 20-μm-thick gas-permeable 35 mm diameter Lumox™ culture plate, the surface of the Lumox is where the cells are cultured. The structures on the device serve as the walls and ceiling of the habitats, creating an enclosed, interconnected space. Three experiments can be run simultaneously in a stainless steel three-well sample plate.

The components of the experiment are illustrated in Fig. 4 (Lin et al., 2017). The entire three-well sample plate is set on the motorized stage of an inverted microscope for long-term (three-week) evolution experiments. Two pairs of syringes with different concentrations of chemicals, e.g., drugs, nutrients, or cytokines, are used to pump fresh medium around the channels. With an optimization in chip design and fabrication processes to allow for mass production at low cost, the throughput of analyzing drugs which rapidly lead to resistant clones under conditions of heterogeneous stress and fragmented cell populations can be increased to analyze dozens or even hundreds of cancer cell lines and drugs in under a month. Combined with deep sequencing, we should be able to obtain a list of causal mutations reflecting diverse tumor contexts of many cell lines efficiently. Compilation of mutations with respect to cancer types and primary chemotherapeutic reagents that had been applied could lead to a large-scale, genetically based profiling of resistance mechanisms. This should represent a database for high-throughput mechanistic studies, suggest drugs that should be avoided for further studies, and provide molecular guidance for avoiding resistance in the first place. Our technology allows high-magnification ($40 \times$) images of the individual cells as a function of position within the drug gradient. The cells are vastly more heterogeneous in morphology than would be expected in conventional assays, and reveals the power of this innovative technology to unmask the much more disturbing (heterokaryon emergence) behavior of cancer cells at the timberline of maximal survivable stress.

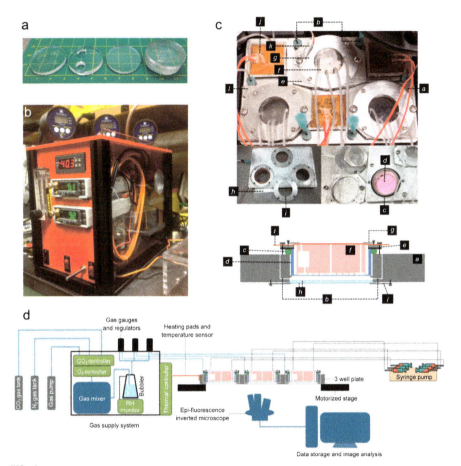

FIG. 4

(A) The PDMS device. From left to right: the patterned layer, the reservoir layer, the capping layer, and the evolution accelerator device. The three stacks of PDMS layers were bonded by oxygen plasma treatment. The reservoirs can trap the bubbles in the tubings and thus guarantee stable fluidic dynamics. (B) The gas supply system. (C) The components of the customized three-well sample plate. (*a*), the main body of three-well plate; (*b*), a pair of gas channels that allow conditioned air to be pumped in and vented out through the septa at the entrance of the gas channels; (*c*), an O-ring designed to seal the space between the well plate and the Lumox™ dish so that it's air-tight; (*d*), the 3 mm Lumox™ dish; (*e*), the Lumox™ dish holder; (*f*), the PDMS device; (*g*), the PDMS chip holders; (*h*), the double layer 35 mm glass windows designed to maintain thermal isolation and prevent water condensation; (*i*), glass window holder; (*j*), heating pads; (*k*), temperature sensor; (*l*), the Microseal® B Adhesive Sealer that keeps the chip from drying out. (D) The setup of the experiment, including the gas supply system, gas channels connection, the medium supply connection and the imaging system.

4.3 APPLICATIONS TO BIOFUELS

The emphasis so far has been the emergence of antibiotic and chemotherapy resistance, but accelerated evolution can also be used for things other than human health. Replacing non-renewable energy sources is one of the biggest and most exciting challenges of our generation. Algae and bacteria are poised to become major renewable biofuels if strains can be developed that provide a high, consistent and robust yield of oil. But currently available bioengineering methods have proven quite slow and expensive for this task. To address this problem, we propose that the technology discussed above offers a new approach to genetic engineering that applies the principles of ecology at the microscale, and dramatically increases the rate at which we can develop new genetic strains. A target of this approach would be to seek out new strains of green algae and bacteria with significantly increased oil production and oil tolerance.

For motile organisms like *E. coli*, that have previously been modified to produce biofuels, we will engineer ecologies to make them more resistant to the biofuels by using the same approach we used to make them resistant to antibiotics in a few hours—replacing the antibiotic with a biofuel like *n*-butanol. The lack of tolerance is what often limits biofuel productivity in these organisms. This procedure can be done with strains we develop, or with strains that have been modified by others. Once modified for higher resistance to biofuel, standard genetic engineering techniques will be used to insert optimized plasmids for biofuel production.

For non-motile organisms, green algae look very promising. We pro- pose working with at least three non-motile algal species—*Neochloris oleoabundans*, *Scenedesmus* sp. and *Botryococcus braunii*. Recent work has shown that neutral lipid production is a metabolic response to starvation—as much as 50% by body weight (Nielsen et al., 2009). Under good light conditions green algae fix carbondioxide and multiply. However, if they are starved (of nitrogen, phosphorus), but still exposed to light—instead of reproducing, they store the excess carbon in the form of neutral lipids. These lipids serve as food if lighting conditions deteriorate, and also protect against intense light conditions. We think that the kind of accelerated evolution presented here could be used with appropriate stress conditions to find mutants that are efficient at this process of lipid production.

The technology we adopt will depend on the organism we work with: some single cellular green algal species accumulate neutral lipids naturally, but are non-motile. Other single cellular organisms like *E. coli* are genetically modified using genes derived from other species to obtain biofuels. The technology we propose requires populations to move from one habitat to another—either naturally or via artificial means in a programmed manner, and the design needs to be tuned to the motility of the organism.

5 CONCLUSIONS

Our cell culture setup in connected microenvironments represents a compelling alternative to traditional microbiological culture methods (e.g., in dishes or flasks) to dissect the complex physiology and behavior of bacteria in real time. Furthermore, such microfluidic device can be adapted to various cell types including cancer cells, and provides a powerful framework for exploring rapid evolution to drug resistance (Han et al., 2016; Lambert et al., 2011). Further, the technology could be used for directing evolution of the powerful bacterial metabolic tools for things such as biofuel production, amongst other things.

REFERENCES

Adler, J. (1975). Chemotaxis in bacteria. *Annual Review of Biochemistry, 44*(1), 341–356.

Adler, M., & Groisman, A. (2012). Microfluidic devices for experiments on bacterial aerotaxis and for high-resolution imaging of motile bacteria. *Biophysical Journal, 102*(3), 151a.

Aizel, K., Clark, A., Simon, A., Geraldo, S., Funfak, A., Vargas, P., et al. (2017). A tuneable microfluidic system for long duration chemotaxis experiments in a 3D collagen matrix. *Lab on a Chip, 17*(22), 3851–3861.

Amselem, G., Guermonprez, C., Drogue, B., Michelin, S., & Baroud, C. (2016). Universal microfluidic platform for bioassays in anchored droplets. *Lab on a Chip, 16*(21), 4200–4211.

Berg, H. (1975). Chemotaxis in bacteria. *Annual Review of Biophysics and Bioengineering, 4*(1), 119–136.

Borer, B., Tecon, R., & Or, D. (2018). Spatial organization of bacterial populations in response to oxygen and carbon counter-gradients in pore networks. *Nature Communications, 9*(1), 769.

Bos, J., Zhang, Q., Vyawahare, S., Rogers, E., Rosenberg, S. M., & Austin, R. H. (2015). Emergence of antibiotic resistance from multinucleated bacterial filaments. *Proceedings of the National Academy of Sciences of the United States of America, 112*(1), 178–183.

Coyte, K., Tabuteau, H., Gaffney, E., Foster, K., & Durham, W. (2016). Microbial competition in porous environments can select against rapid biofilm growth. *Proceedings of the National Academy of Sciences of the United States of America, 114*(2), E161–E170.

Crooks, J., Stilwell, M., Oliver, P., Zhong, Z., & Weibel, D. (2015). Decoding the chemical language of motile bacteria by using high-throughput microfluidic assays. *Chembiochem, 16*(15), 2151–2155.

Davis, R., Markham, A., & Balfour, J. A. (1996). Ciprofloxacin. An updated review of its pharmacology, therapeutic efficacy and tolerability. *Drugs, 51*(6), 1019–1074.

Fenchel, T., & Finlay, B. (2008). Oxygen and the spatial structure of microbial communities. *Biological Reviews, 83*, 553–569.

Flemming, H. C., Wingender, J., Szewzyk, U., Steinberg, P., Rice, S. A., & Kjelleberg, S. (2016). Biofilms: An emergent form of bacterial life. *Nature Reviews. Microbiology, 14*(9), 563–575.

Good, B. H., McDonald, M. J., Barrick, J. E., Lenski, R. E., & Desai, M. M. (2017). The dynamics of molecular evolution over 60,000 generations. *Nature, 551*(7678), 45–50.

Han, J., Jun, Y., Kim, S. H., Hoang, H. H., Jung, Y., Kim, S., et al. (2016). Rapid emergence and mechanisms of resistance by U87 glioblastoma cells to doxorubicin in an in vitro tumor microfluidic ecology. *Proceedings of the National Academy of Sciences of the United States of America, 113*(50), 14283–14288.

Hol, F. J., & Dekker, C. (2014). Zooming in to see the bigger picture: Microfluidic and nanofabrication tools to study bacteria. *Science, 346*(6208), 1251821.

Hu, B., & Tu, Y. (2013). Precision sensing by two opposing gradient sensors: How does Escherichia coli find its preferred pH level? *Biophysical Journal, 105*(1), 276–285.

Jasovsky, D., Littmann, J., Zorzet, A., & Cars, O. (2016). Antimicrobial resistance—A threat to the world's sustainable development. *Upsala Journal of Medical Sciences, 121*(3), 159–164.

Jiang, L., Ouyang, Q., & Tu, Y. (2009). A mechanism for precision-sensing via a gradient-sensing pathway: A model of Escherichia coli thermotaxis. *Biophysical Journal, 97*(1), 74–82.

Keymer, J. E., Galajda, P., Muldoon, C., Park, S., & Austin, R. H. (2006). Bacterial metapopulations in nanofabricated landscapes. *Proceedings of the National Academy of Sciences of the United States of America, 103*(46), 17290–17295.

Khan, S., Spudich, J., McCray, J., & Trentham, D. (1995). Chemotactic signal integration in bacteria. *Proceedings of the National Academy of Sciences of the United States of America, 92*(21), 9757–9761.

Lambert, G., Estevez-Salmeron, L., Oh, S., Liao, D., Emerson, B. M., Tlsty, T. D., et al. (2011). An analogy between the evolution of drug resistance in bacterial communities and malignant tissues. *Nature Reviews Cancer, 11*(5), 375–382.

Lambert, G., Liao, D., & Austin, R. H. (2010). Collective escape of chemotactic swimmers through microscopic ratchets. *Physical Review Letters, 104*(16), 168102.

Lebeaux, D., Ghigo, J. M., & Beloin, C. (2014). Biofilm-related infections: Bridging the gap between clinical management and fundamental aspects of recalcitrance toward antibiotics. *Microbiology and Molecular Biology Reviews, 78*(3), 510–543.

Lin, K-C., Torga, G., Wu, A., Rabinowitz, J. D., Murray, W., Sturm, J. C., et al. (2017). Epithelial and mesenchymal prostate cancer cell population dynamics on a complex drug landscape. *Convergent Science Physical Oncology, 3*(4), 045001.

Maeda, K., Imae, Y., Shioi, J. I., & Oosawa, F. (1976). Effect of temperature on motility and chemotaxis of Escherichia coli. *Journal of Bacteriology, 127,* 1039–1046.

Morris, R., Phan, T., Black, M., Lin, K., Kevrekidis, I., Bos, J., et al. (2017). Bacterial population solitary waves can defeat rings of funnels. *New Journal of Physics, 19*(3), p. 035002.

Nielsen, D. R., Leonard, E., Yoon, S. H., Tseng, H. C., Yuan, C., & Prather, K. L. (2009). Engineering alternative butanol production platforms in heterologous bacteria. *Metabolic Engineering, 11*(4–5), 262–273.

Paster, E., & Ryu, W. (2008). The thermal impulse response of Escherichia coli. *Proceedings of the National Academy of Sciences of the United States of America, 105*(14), 5373–5377.

Salman, H., & Libchaber, A. (2007). A concentration-dependent switch in the bacterial response to temperature. *Nature Cell Biology, 9*(9), 1098–1100.

Taylor, B., Zhulin, I., & Johnson, M. (1999). Aerotaxis and other energy-sensing behavior in Bacteria. *Annual Review of Microbiology, 53*(1), 103–128.

Weibel, D. B., Diluzio, W. R., & Whitesides, G. M. (2007). Microfabrication meets microbiology. *Nature Reviews. Microbiology, 5*(3), 209–218.

Wessel, A. K., Hmelo, L., Parsek, M. R., & Whiteley, M. (2013). Going local: Technologies for exploring bacterial microenvironments. *Nature Reviews. Microbiology, 11*(5), 337–348.

Yang, Y., & Sourjik, V. (2012). Opposite responses by different chemoreceptors set a tunable preference point in Escherichia coli pH taxis. *Molecular Microbiology, 86*(6), 1482–1489.

Yu, H. (2002). Aerotactic responses in bacteria to photoreleased oxygen. *FEMS Microbiology Letters, 217*(2), 237–242.

Zhang, Q., Bos, J., Tarnopolskiy, G., Sturm, J., Kim, H., Pourmand, N., et al. (2014). You cannot tell a book by looking at the cover: Cryptic complexity in bacterial evolution. *Biomicrofluidics, 8*(5), p. 052004.

Zhang, Q., Lambert, G., Liao, D., Kim, H., Robin, K., Tung, C. K., et al. (2011). Acceleration of emergence of bacterial antibiotic resistance in connected microenvironments. *Science, 333*(6050), 1764–1767.

Zhang, Q., Robin, K., Liao, D., Lambert, G., & Austin, R. H. (2011). The Goldilocks principle and antibiotic resistance in bacteria. *Molecular Pharmaceutics, 8*(6), 2063–2068. https://doi.org/10.1021/mp200274r.

FURTHER READING

Wang, P., Zhang, X. N., Wang, L., Zhen, Z., Tang, M. L., & Li, J. B. (2010). Subinhibitory concentrations of ciprofloxacin induce SOS response and mutations of antibiotic resistance in bacteria. *Annals of Microbiology, 60*(3), 511–517.

Wickens, H. J., Pinney, R. J., Mason, D. J., & Gant, V. A. (2000). Flow cytometric investigation of filamentation, membrane patency, and membrane potential in Escherichia coli following ciprofloxacin exposure. *Antimicrobial Agents and Chemotherapy, 44*(3), 682–687.

A microfluidic fluidized bed to capture, amplify and detect bacteria from raw samples

Lucile Alexandre[*,†,‡,a], **Iago Pereiro**[*,†,‡,a], **Amel Bendali**[*,†,‡], **Sanae Tabnaoui**[*,†],
Jana Srbova[§], **Zuzana Bilkova**[§], **Shane Deegan**[¶,‖], **Lokesh Joshi**[¶,‖],
Jean-Louis Viovy[*,†,‡], **Laurent Malaquin**[*,†,‡], **Bruno Dupuy**[#],
Stéphanie Descroix[*,†,‡,1]

*Laboratoire Phyisico Chimie Curie, Institut Curie, PSL Research University, Paris, France
†Sorbonne Universités, UPMC Univ Paris 06, Paris, France
‡Institut Pierre-Gilles de Gennes, Paris, France
§Department of Biological and Biochemical Sciences, Faculty of Chemical Technology, University of Pardubice, Pardubice, Czech Republic
¶Aquila Bioscience limited, Business Innovation Centre, National University of Ireland Galway, Galway, Ireland
‖Glycoscience Group, National Centre for Biomedical Engineering Science, National University of Ireland Galway, Galway, Ireland
#Laboratory Pathogenesis of Bacterial Anaerobes, Department of Microbiology, Institut Pasteur, Paris, France
1Corresponding author: e-mail address: stephanie.descroix@curie.fr

CHAPTER OUTLINE

---

[a]Both authors contribute equal to this work (co-first authors).

Methods in Cell Biology, Volume 147, ISSN 0091-679X, https://doi.org/10.1016/bs.mcb.2018.07.001

Abstract

Bacterial contamination and subsequent infections are a major threat to human health. An early detection in the food chain, clinics or the environment, is key to limit this threat. We present a new concept to develop low-cost hand-held devices for the ultra-sensitive and specific detection of bacteria in a one-step process of 2–8 h, directly from complex raw samples. This approach is based on a novel microfluidic magnetic fluidized bed. It reaches a 4 CFU (colony forming unit) sensitivity with high quantification accuracy in a large dynamic range of $100–10^7$ CFU/mL. The versatility of the approach was demonstrated with the detection of different bacteria strains, among which Salmonella Typhimurium and *E. coli* O157:H15. Additionally, the method is sensitive to infectious bacteria only, a criterion requested by main applications and currently requiring additional culture steps of one to several days.

1 INTRODUCTION

Infectious disease are considered as a major global threat to human health by the Worlds Health Organization (2014). Despite the progress in hygiene and antibiotics development, there is still a crucial need for innovative approaches to detect and quantify the presence of pathogen in a wide range of domains such as clinics, environment or food safety chains. The current gold standard for bacteria detection relies on plating and colony counting. Although highly efficient, this approach is time consuming (several days may be needed depending on detection accuracy) and requires skilled personnel that perform the manual labor. To overcome these limitations, different strategies have been proposed. Among them, some are based on antibody targeting (ELISA or immunoprecipitation) (GutiéRrez et al., 1997; Wen et al., 2013), flow cytometry (Karo et al., 2008) or PCR (Pathmanathan et al., 2003). In parallel, due to its unique capabilities in hydrodynamic control, miniaturization and automation, microfluidics has emerged as a high performance technology for different applications in bioanalysis (Hernández-Neuta et al., 2018), including microdevices for bacteria analysis (Hou, Bhattacharyya, Hung, & Han, 2015; Wu, Willing, Bjerketorp, Jansson, & Hjort, 2009). However, these methods often require pre-enrichment steps to reach the expected sensitivity for food analysis.

Here, we propose a new microfluidic approach based on miniaturized magnetic fluidization technology. A fluidized bed is generated when particles are suspended in a steady-state dynamic regime, with viscous drag and gravity as counterbalancing force. In the macroscale, this yields efficient chaotic stirring and enhances the heat and mass transfer between the solid and liquid phases, as compared to static packed beds with percolating fluids. We have adapted the concept of the fluidized bed to the microscale, replacing gravitational forces by magnetic forces as the counterbalance to fluid drag. This is possible since: (i) gravitational forces are several orders of magnitude weaker than drag forces in our system and (ii) magnetic micrometric beads and external permanent magnets were used as the solid phase and the magnetic field source, respectively. The design of the microfluidic device was optimized in order to compensate for the decaying intensity of the magnetic field and maintain an approximate balance between the hydrodynamic drag and magnetic forces in all channel sections (Pereiro, 2016). Our approach is well suited for bacteria immunoextraction, as it ensures a high bead density and specific surface to favor pathogen capture, combined the system's low working backpressure and high resistance to clogging, necessary for complex sample analysis. Here, we demonstrate that this approach can be used to capture different pathogens with high efficiency, demonstrated in particular for Salmonella enterica serovar Typhimurium. Once the bacteria are captured, growth medium can be percolated through the solid phase to multiply in situ the live bacteria. The volume occupied by the newly formed bacteria leads to geometrical modifications of the fluidized bed that can be monitored to obtain a specific quantification of the initial number of bacteria present in the sample. Thus, all the required steps are performed in a single microfluidic chamber on a previously introduced solid phase, enabling a one-step capture, amplification and detection process.

2 METHOD

We describe here an innovative method to locally extract and pre-concentrate bacteria from a liquid phase. The method includes the detection and evaluation of the concentration of bacteria in the tested sample, performed directly inside the system.

2.1 PRESENTATION OF THE MICROFLUIDIZED BED

In a fluidized bed, a phase of solid particles is under the influence of two opposing forces: an upward drag force and a downward gravitational force. In the right range of fluid flow, a chaotic hydrodynamic regime establishes, in which the behavior of the solid phase is in many ways comparable to the properties of a fluid. Fluidized beds are used in industry to enhance the contact in solid/liquid mixtures, enhancing surface reactions and mass and temperature transfers as compared to more static systems. We have developed a micro-fluidized bed (MFB) based on the idea of macroscale fluidized bed, in which a global equilibrium is established between two forces: (i) the drag force created by the fluid percolating through the solid phase inside the micro-chamber of the system and (ii) a magnetic force applied on a solid phase of magnetic microparticles, the result of the presence of a magnetic field created by an external magnet (Fig. 1). Because of the reduced dimensions of the microfluidic chip, the obtained hydrodynamic regime is inherently of low Reynolds number and, hence, a purely laminar flow is obtained. Therefore, the mixing between the solid and liquid phases is not obtained through a highly turbulent regime (as is the case of macroscale fluidized beds), but through the constant recirculation of the solid phase inside the microfluidic chamber, which is obtained with a specific height and tapered geometry (Pereiro, 2016). Briefly, particles are accelerated by a strong drag force near the chamber entrance and moved toward the downstream front of the solid phase. Magnetic and drag forces equilibrate at this point, but the particles present in this region are constantly displaced toward the sides of the chamber due to the arrival of other particles. Near the wall of the chamber, due to the local non-slipping condition, drag forces weaken and magnetic forces become predominant, bringing the particles back to the entrance. Because of this recirculation, the entire solid phase is

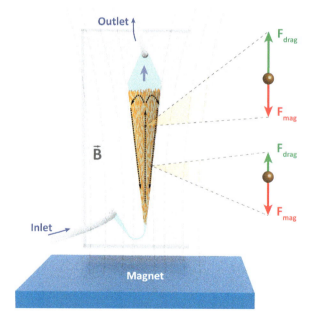

FIG. 1

Schematic of the fluidized bed with a representation of local balances of forces applied on the magnetic beads. The beads are acted by opposing drag forces, due to the flow of liquid percolating through the solid phase, and magnetic forces, due to the gradient of magnetic field generated by the permanent magnet. The recirculation of the beads inside the matrix is represented by black arrows.

in constant movement, maximizing the exposure of the particle surface to the percolating liquid and avoiding the formation of stagnant areas. Importantly, the solid particles present in the chamber do not act as individual entities, but form particle chains due to their dipole to dipole interaction in the presence of a magnetic field. These chains are constantly restructured when reaching the chamber inlet due to high local shear stresses. This restructuring ensures the constant renewal of the surface in contact with the liquid phase. Finally, due to its dynamic nature, the solid phase adapts to the entrance of large interfering particles that could be present in the liquid phase. Because they are non-magnetic, these particles are unaffected by the magnetic field and directly dragged to the outlet, avoiding the clogging of the system or the stacking of incoming solids.

2.2 CHIP DESIGN

To obtain an adequate recirculation of the solid phase, the design of the chip must answer specific criteria:

- The main chamber has a v-shape with an entrance angle of 13°. In particular, our chamber is fabricated with a diamond shape 13 mm long and 2.6 mm of maximum width (chamber volume of 3 µL). The outlet is positioned at the extreme end of the chamber opposite the 13° angle.
- An inlet channel (100 µm width) in shape of an elbow between the entrance and the main chamber avoids beads from flowing upstream when the fluid flow is stopped due to geometrical particle constriction.
- The height of the chamber is 50 µm to limit the size of the particle chains (Pereiro, 2016) (Fig. 2).

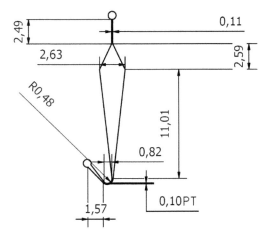

FIG. 2

Blueprint of the design of fluidized bed (all numbers are in millimeter, *R* means radius and PT means Point of Tangency). This blueprint is used to micro-milled the design on brass mold with a height of 50 µm.

2.3 CHIP FABRICATION

2.3.1 Materials section

Molds for PDMS casting can be fabricated by several techniques, including micro-milling and photolithography. The mold will contain the chip design previously described in a mirror geometry.

The choice of the chip material is dependent to the application. For the experiments presented, we used polydimethylsiloxane (PDMS, Sylgard184, Dow Corning).

2.3.2 Fabrication of PDMS chips

(a) Mix PDMS with its crosslinker (at a ratio of 1:10).

(b) Pour the PDMS on the brass, previously surrounded by thermal tape for PDMS retention. The layer should have an approximate thickness of 3–5 mm. Leave it recirculate at 60° for 2 h. This will become the upper part of the chip.

(c) Pour 12 g of the same mixed PDMS on a flat Petri dish 9 cm in diameter (or equivalent dish size to PDMS mass) to form a thin layer of 2 mm. To ensure a uniform depth of the layer, leave the Petri dish reticulate overnight on a horizontal surface (e.g., optical table).

(d) Remove both PDMS layers from mold and dish with the aid of metallic tweezers.

(e) Perforate the cast part in the two extreme ends of the chamber with a PDMS puncher to create the inlet and outlet (0.75 or 1 mm in diameter).

(f) Carefully clean both parts with isopropanol and dry with air gun.

(g) Insert them in an oxygen plasma cleaner and treat for 30 s for surface activation.

(h) Put both parts in contact for instant bonding and chamber sealing.

(i) (optional) A surface treatment can be introduced in the chamber to minimize the adsorption of sample molecules and particles on the inner walls of the chip. For example, a water solution of poly-dimethyl-acrylamide bearing epoxy polydimethylacrylamideco-allyl glycidyl ether (PDMA-AGE) 0.5% (*w/v*) can be injected through the inlet of the system with the aid of a syringe and left incubate for 1 h. After this time, the solution is washed away by introducing an excess of distilled water through the inlet, and the chip is dried in an oven at 60 °C for 1 h.

2.4 SET UP

The necessary set up for fluidic and thermal control (Fig. 3), as well as visualization, is described:

- A 1.5 T permanent magnet (e.g., NdFeB magnet N50, Chen yang Technologies, $2 \times 2 \times 3$ cm) is positioned horizontally with one of the poles facing the chip and aligned with the main axis of the microfluidic chamber, 2 mm from the chamber entrance.

- Control of the flow is most convenient with a pressure-based system for reservoir change. For e.g., a MAESFLO controller (Fluigent) adjusts the air pressure inside the entrance reservoir and a flow sensor (Flow-unit) can be connected to

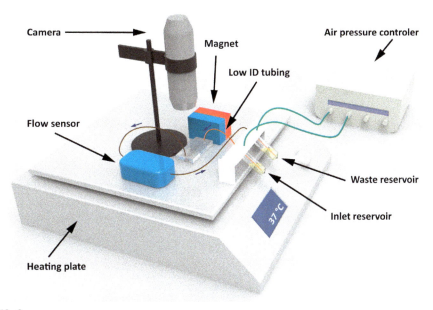

Camera

Magnet

Low ID tubing

Air pressure controler

Flow sensor

Waste reservoir

37°C

Inlet reservoir

Heating plate

FIG. 3

Schematic representation of the set-up of the fluidized bed with a USB camera positioned on top of the chip, a heating plate to keep the system at 37°C, a permanent magnet near the entrance of the chip, a tubing with low ID at the entrance of the chip, a flow sensor at the output, a pressure control to inject the liquid in the inlet reservoir, and a waste reservoir to collect the outflowing liquid.

the outlet for flow-rate reading. The reading from the flow sensor is used by the Fluigent MAESFLO software as a feedback to regulate the inlet pressure and obtain the desired flow-rate.

- Temperature is necessary for detection/incubation experiments. For this, the PDMS chip can be positioned on an indium tin oxide (ITO) glass slide (for e.g., for detailed experimental visualization of bacteria through an inverse microscope) or on a heating plate (when only solid phase tracking is needed, for normal bacteria detection and quantification). In the case of using an ITO system, a thermal probe can be inserted between the bound parts of the PDMS chip, close to the chamber, for direct temperature measurement in the experimental area. A thermal controller (e.g., Eurotherm 3508) can be used in this case to heat the ITO plate regulated by the feedback reading of the probe.
- High chemical resistance and low adhesion PEEK tubings are recommended for the fluidic path (between the chip, the flow unit and the outlet waste reservoir) width and external diameters of 1/32 in as a good fit for 0.75 mm holes punched in the PDMS. For the tubing at the entrance of the chip, a small length and internal diameter (e.g., <15 mm and <100 μm, respectively) define an adequate fluidic resistance of the system and minimize the connecting volume between reservoir and chip.

- A simple USB camera can be positioned on top of the chip to track changes in the size of the solid phase. The use of microscopy is only necessary to study the presence/propagation of bacterial units.
- To prevent the inefficient washing of uncaptured bacteria, valves with internal dead-volumes are best avoided. If used, particularly to obtain automation, models with a minimal internal fluidic path are recommended.

2.5 INSERTION OF BEADS INSIDE THE CHIP

2.5.1 Material section

Magnetic microbeads with diameters in a 1–5 μm range can be used as the dynamic solid phase. For example, Dynabeads™ anti-Salmonella (Thermofisher Scientific, 71002) (2.8 μm in diameter) were previously used for the specific capture of Salmonella strains (Pereiro et al., 2017).

2.5.2 Protocol for the bead insertion inside the MFB (*Fig. 4*)

(a) Plug the whole fluidic path and fill it with the buffer solution, including the PDMS chip (ensuring no remaining bubbles in the microfluidic chamber).
(b) Detach the inlet tubing from the chip and replace it by a pipette tip. Push liquid from the outlet to partially fill the tip.
(c) Introduce the magnetic beads in the partially filled tip by pipetting. The beads are previously washed three times in buffer.
(d) Guide the magnetic beads toward the chip with the aid of a small magnet, and further through the entrance elbow to the main chamber.
(e) When all beads are inside the main chamber, remove the pipette tip and replace it with the original inlet tubing. A drop to drop contact is recommended to avoid bubbles. The magnet should stay in place to avoid that beads escape the chamber during operation.
(f) Adjust the magnet at 2 mm of the elbow channel and align it with the axis of the microfluidic chamber. When the buffer solution is flowed from the inlet to the outlet of the main chamber, the magnetic beads homogenize because of the created recirculation. The flow can be stopped at any time, at which point the beads will move toward the elbow and create a compact ensemble.

2.6 CULTURE OF BACTERIA

2.6.1 Materials section

Glycerol Anhydrous was purchased at Sigma-Aldrich® (56-81-5), as agar (9002-18-0) and LB-broth (L2542). Inoculating loop (Sigma-Aldrich®, I8263) and Lazy-L spreader (Sigma-Aldrich®, Z376779) can also help the preparation of bacteria. Culture dishes were produced in Curie Institute.

Bacteria can be stored as aliquots at −80 °C in liquid form (glycerol solution) or on culture dishes.

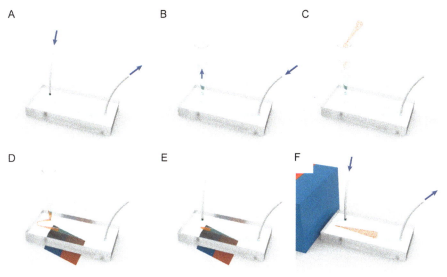

FIG. 4

Schematic representation of the insertion of magnetic beads inside the main chamber of the PDMS chip. The fluidic path is filled with buffer (A) and the inlet tubing is detached to insert a pipette tip (B) where the solution of beads is dispensed (C). The beads are guided inside the chamber by a magnet (D) up to the entrance. The pipette tip is removed and replaced by the inlet tubing (E) and the magnet is positioned at 2 mm of the elbow channel of the PDMS chip and aligned with the chamber (F).

2.6.2 Protocol for glycerol stocks

(a) Prepare a solution of 50% glycerol diluted in sterile ultra-pure water (ddH$_2$O)

(b) Prepare an overnight bacterial culture (Protocol 1.7 A, B, C) in nutritional medium such as LB-broth or LB-agar.

(c) Mix bacterial culture and glycerol solution at a 1:1 ratio.

(d) Divide the final solution in aliquots and transfer directly to a −80°C fridge.

2.6.3 Protocol for solid culture on culture dishes

(a) Dip an inoculating loop in an aliquot to collect some bacteria sample.

(b) Spread it on 1/3 of the surface of a culture dish.

(c) Incubate for 24 h at 37°C.

(d) Spread with a single linear movement some of the colonies that have developed during the night another third of the culture dish.

(e) Incubate for 24 h at 37°C.

(f) Spread with a single linear movement some of the colonies that have developed during the night another third of the culture dish.

(g) Collect one single colony to create a liquid culture (Protocol 1.7).

Bacteria can be stored long term at −80°C. It's also possible to keep them on culture dish at 37°C for a few days or in the fridge for a few weeks.

2.7 EVALUATION OF CAPTURE EFFICIENCY OF MFB

The method can be applied to many bacterial strains that can be captured through specific membrane receptors and that multiply in the presence of liquid nutritional medium. However, not all bacterial strains can be captured efficiently with known ligands. A sufficient capture efficiency is necessary to (i) ensure that enough bacteria is extracted from the original sample and (ii) that during the incubation step the multiplying bacteria are recaptured by the solid phase instead of being released downstream. Based on previous experiments, a capture rate of at least ~20% is necessary to ensure that detection is viable.

Protocol to establish capture efficiency of chosen strain (Fig. 5):

A. With an inoculating loop take a small bacteria sample from an aliquot stored at $-80\,°C$ or from a recently cultured dish.

B. Dip the loop into a 50 mL falcon with 20 mL of LB-broth or alternative nutritional medium for the chosen bacterial strain.

C. Incubated the falcon overnight at $37\,°C$ and 244 rpm (these are typical good conditions for strains such as Salmonella, other strains might have more adequate incubating temperatures and times). The next day, mix 2 mL of this solution to 18 mL of LB and let it incubate for two additional hours.

D. Check the concentration of bacteria in the solution by optical density (e.g., Ultrospec 10 Amersham Biosciences). Typically, a reading of 0.9 corresponds to a concentration of bacteria of 10^9 CFU.

E. Dilute in the desired buffer until the concentration ensures that the maximum number of captured bacteria will be countable (200–700 CFU is considered the safest range for quantification).

F. Inject a sample volume of 50 μL through the chip inlet at a flow-rate of 1 μL/min. Optionally, the sample can be collected at the outlet directly with a reservoir or by placing a microfluidic tubing of enough volume from which the liquid is collected after the step.

G. Wash the beads inside the chip with 40 μL of at a flow-rate of 1.5 μL/min. Optionally, the washing liquid can be collected at the outlet.

H. Remove the permanent magnet and keep a flow rate of washing buffer (>1 μL/min) to collect the magnetic beads at the outlet.

I. For capture quantification, plate the following liquids on culture dishes:
   a. 50 μL of initial sample reservoir.
   b. The total volume of sample and washing buffer percolated through the system (optional).
   c. The beads flushed from the chip after the capture step.

J. Count the colonies after an overnight incubation at $37\,°C$. The capture rate is calculated as the ratio between captured bacteria [1] (cultured beads) and initial sample [3]. Alternatively, the captured bacteria [1] can be compared to the non-captured bacteria [2]. This last method generally results in slightly more reproducible results, as the bacteria counted are directly those used during the experiment.

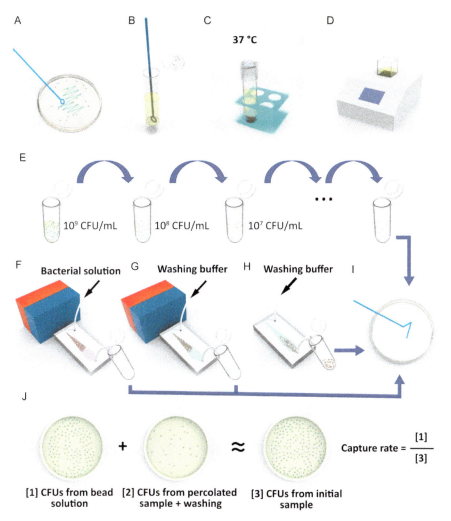

FIG. 5

Schematic representation of the protocol to evaluate the capture efficiency of bacteria in situ by: (A) and (B) inoculation of bacteria inside a nutritional medium; incubation period of 37 °C (C); control of concentration by optical density (D). The wanted concentration is obtained with dilutions (E). Once the sample is prepared, 50 μL is injected inside the matrix of magnetic beads at 1 μL/min at RT (F) with a washing step at 1.5 μL/min of 40 μL of buffer (H). The sample, the washing buffer and the beads can be collected and plated on culture dishes (I). Counting the colonies on the culture dishes after an overnight incubation gives the capture rate of the system (J).

2.8 CALIBRATION OF THE SYSTEM FOR DETECTION OF BACTERIA

The detection of bacteria is obtained by the visual observation of the expansion of the solid phase, obtained by letting the bacteria multiply in a nutritious medium inside the chip. The newly formed bacteria will occupy an ever increase volume that will affect the fluidization equilibrium of the bed (Pereiro et al., 2017). As each bacterial strain presents unique doubling times, membrane volume and capture affinity to the chosen ligand, the times of expansion depending on the initial concentration of the sample must be initially studied:

(a) Follow the steps from the previous section up to G (including sample capture and washing; Fig. 6A and B). In step E choose adequate dilutions to test the desired concentration of bacteria in the sample. The final dilution is obtained by spiking the bacteria in the sample matrix of interest. First, the capture (Fig. 6A) and the washing (Fig. 6B) steps were made as described previously (up to G on Fig. 5).

(b) Flow nutritional medium through the bed of beads at 150 nL/min (Fig. 6C) and set an adequate incubation temperature. Each strain multiplies at different optimal medium and temperature (e.g., for Salmonella strains, LB-broth and 37 °C).

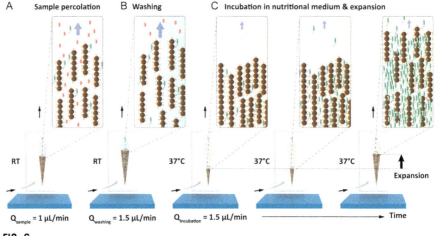

FIG. 6

Schematic representation of a bacteria detection experiment: percolation of the sample inside the beads (A), washing step with buffer (B) and in situ incubation of bacteria in a nutritional medium at 37 °C (C). The growth of bacteria inside the matrix of magnetic beads and the expansion due to increased volume occupancy are represented as a function of time.

(c) The bed expands after a given time that can range from several minutes to several hours depending on initial sample concentration, receptor–ligand affinity, multiplication speed and volume of entity. Measure this time following the description of the next section.

(d) After sufficient expansion (>200 µm) decontaminate the system with an antibacterial solution and dispose the chip.

2.9 MEASURE OF MFB EXPANSION

2.9.1 Materials section

Camera needed is a simple camera such as a USB Dino-Lite camera (Dino-Lite Digital Microscope). Software used was Image J however this protocol can be adapted to any processing software.

2.9.2 Protocol to measure the fluidized bed expansion

(a) Set the camera to take images of the bed with a frequency of at least 1 image/min.

(b) After recording is complete, use a processing software (e.g., ImageJ) to set the geometrical scale of the images.

(c) With the first images, before any bed expansion, establish the base zero position of the bed after incubation was started (the bed front stays stable before any expansion due to bacteria).

(d) Track the front position of the bed for subsequent images (either manually or by creating a script that recognizes the change in the local intensity value) (Fig. 7A).

(e) 200 µm of expansion from the zero base are considered a positive detection of bacteria.

Generally, the correlation is linear between the time of expansion and the logarithm of the number of bacterial entities in the initial sample (Fig. 7A).

Therefore, the MFB system can be used to evaluate the number of bacteria present in a sample with a good degree of confidence. This requires repeating this calibration steps for several initial concentrations of bacteria allows obtaining a specific graph for the bacteria/ligand support that can be used as a calibration curve.

Once obtained, samples with bacteria at unknown concentrations can be tested for both revealing their presence and quantifying their concentration.

The calibration curve can be used to establish the maximum time matching the time needed for only one bacterium to create a bed expansion. For times longer than this maximum, we can consider that no expansion should be observed, something that would indicate either a weakened strain, an inefficient expansion or the presence of other bacterial strains that non-specifically adsorb to the solid support. For complex samples containing various strains, it is therefore important to repeat this calibration protocol with the strains that should not be detected, to ensure that the ligand support is sufficiently specific.

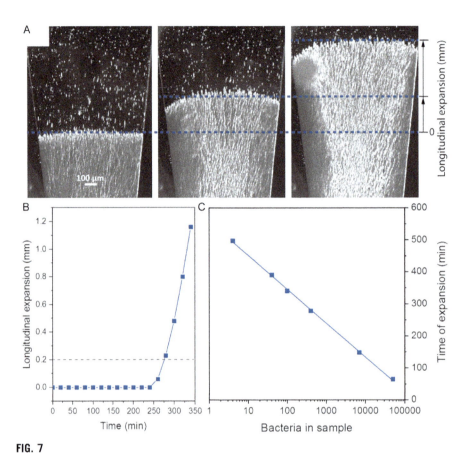

FIG. 7

(A) Images of the expansion of the fluidized bed due to bacteria growth; (B) longitudinal measured expansion of the bed as function of time and (C) time of expansion (for 200 µm) as a function of the bacteria initially present in the sample (Salmonella Typhimurium).

2.10 **TESTING OF ANTIBIOTICS**

As previously indicated, changes in the time of expansion are a sign of changes in the system. If the strain has already been studied in terms of capture and expansion behavior, and is preparation before introduction in the chip follows a standard procedure, as previously described, this changes can be used to study other properties. An example is the presence of a certain concentration of antibiotic in the sample or the nutritious medium to investigate the bacteria susceptibility. This last case is particularly useful to study the efficacy of antibiotics on bacteria starting from well-established initial conditions. For this, a protocol is described below:

(a) Perform all the steps from Section 2.7 up to step G (no collection of liquid during capture and washing steps) for the desired initial concentration of bacteria.

(b) Spike the amount of antibiotic to be tested in the nutritive medium and percolate through the system as indicated in Section 2.8.

(c) Perform image recording as indicated in Section 2.9.

(d) Evaluate the time of expansion. Typically, the antibiotic delays the doubling time, resulting in delayed times of expansion, as expected for the calibration curve previously obtained. If the concentration of antibiotic is sufficiently high no expansion no the bed will take place even after times much longer those expected. This is the critical concentration that avoids strain development.

An example of the delayed times obtained for various concentrations of antibiotic (chloramphenicol) with the incubation of Salmonella Typhimurium is presented in Fig. 8. Because the time of expansion is infinite after a certain antibiotic concentration thershold, the data obtained with the MFB is plotted for the inverse of the time of expansion. As seen, the behavior closely corresponds to the results obtained with optical detection in liquid incubation. The division time in the presence of antibiotics can be calculated knowing the doubling time in standard conditions and accounting for the time delay in each condition (the number of divisions to obtain the 200 µm is always identical) (Fig. 8).

3 DISCUSSION

A microfluidic fluidized bed is applied here to bacteria immunocapture, growth and detection on a single chip. This microfluidic fluidized bed presents interesting features for this application such as the ability to accommodate high flow rates, to improve fluid/solid support contact and to avoid clogging. As a result of these unique features, this fluidized bed device is successfully applied to the capture of different types of bacteria, among which Salmonella Typhimurium. Being able to grow bacteria inside the chip following their capture offers the possibility to precisely quantify the initial amount of live bacteria in the sample just by imaging the fluidized solid phase. Indeed, the bed expansion time during bacteria growth is directly correlated with the initial amount of bacteria present in the sample. Finally taking benefit of the bacteria growth on chip, the device can also be used to investigate bacteria susceptibility toward antibiotics. In particular, the increase of the expansion time is proportional to the susceptibility of the bacteria to the antibiotic. Compared to existing methods, this approach is highly versatile regarding the different functions that can be performed and integrated on a single chip. However, it is based on immunocapture and, thus, it requires that the bacteria to be analyzed have been tested for effective captured by ligand on the solid particles. Besides, contrary to mass spectrometry or sequencing, it is not convenient to identify different unknown bacterial strains from a given sample, as only one strain is targeted at a time.

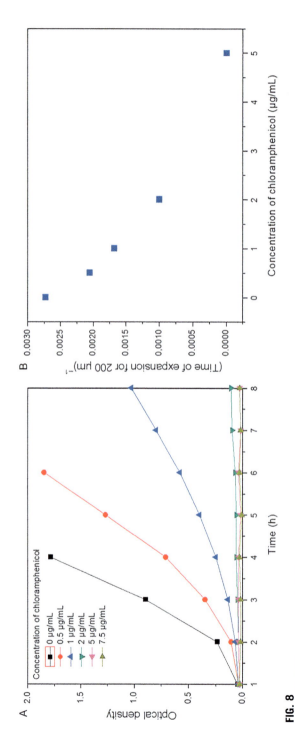

FIG. 8

(A) Exponential growth of Salmonella Typhimurium in the presence of LB-broth and an antibiotic (chloramphenicol) measured by optical density as function of time. The concentration of this antibiotic in the nutritive medium injected during the incubation step also affects the time of expansion of the MFB (B).

Finally, the approach is well suited to detect live bacteria in case of contamination by identified bacteria. Indeed, it allows to capture, grow and detect live bacteria on a single one-chamber chip, with shortened analysis time (<10 h) and with sensitivity at 4 CFU/mL without any pre-amplification step. These features make this microfluidic fluidized bed a competitive approach for the detection of live bacteria present in raw samples such as milk.

REFERENCES

GutiéRrez, R., GonzáLez, I., GarcíA, T., Carrera, E., Sanz, B., HernáNdez, P. E., et al. (1997). Monoclonal antibodies and an indirect ELISA for detection of psychrotrophic bacteria in refrigerated milk. *Journal of Food Protection*, *60*, 23–27.

Hernández-Neuta, I., Pereiro, I., Ahlford, A., Ferraro, D., Zhang, Q., Viovy, J.-L., et al. (2018). Microfluidic magnetic fluidized bed for DNA analysis in continuous flow mode. *Biosensors & Bioelectronics*, *102*, 531–539.

Hou, H. W., Bhattacharyya, R. P., Hung, D. T., & Han, J. (2015). Direct detection and drug-resistance profiling of bacteremias using inertial microfluidics. *Lab on a Chip*, *15*, 2297–2307.

Karo, O., Wahl, A., Nicol, S. B., Brachert, J., Lambrecht, B., Spengler, H. P., et al. (2008). Bacteria detection by flow cytometry. *Clinical Chemistry and Laboratory Medicine: CCLM FESCC*, *46*, 947–953.

Pathmanathan, S. G., Cardona-Castro, N., Sánchez-Jiménez, M. M., Correa-Ochoa, M. M., Puthucheary, S. D., & Thong, K. L. (2003). Simple and rapid detection of Salmonella strains by direct PCR amplification of the hilA gene. *Journal of Medical Microbiology*, *52*, 773–776.

Pereiro, I. (2016). *Microfluidic magnetic fluidized bed for bioanalytical applications*. (phd thesis). Paris VI: Université Pierre et Marie Curie.

Pereiro, I., Bendali, A., Tabnaoui, S., Alexandre, L., Srbova, J., Bilkova, Z., et al. (2017). A new microfluidic approach for the one-step capture, amplification and label-free quantification of bacteria from raw samples. *Chemical Science*, *8*, 1329–1336.

Wen, C.-Y., Hu, J., Zhang, Z.-L., Tian, Z.-Q., Ou, G.-P., Liao, Y.-L., et al. (2013). One-step sensitive detection of Salmonella typhimurium by coupling magnetic capture and fluorescence identification with functional nanospheres. *Analytical Chemistry*, *85*, 1223–1230.

World Health Organization (2014). *Antimicrobial resistance: Global report on surveillance*. World Health Organization.

Wu, Z., Willing, B., Bjerketorp, J., Jansson, J. K., & Hjort, K. (2009). Soft inertial microfluidics for high throughput separation of bacteria from human blood cells. *Lab on a Chip*, *9*, 1193–1199.

Microfluidics for cell migration

Micro-engineered "pillar forests" to study cell migration in complex but controlled 3D environments

5

Jörg Renkawitz, Anne Reversat, Alex Leithner, Jack Merrin, Michael Sixt[1]

Institute of Science and Technology Austria (IST Austria), Klosterneuburg, Austria
[1]Corresponding author: e-mail address: michael.sixt@ist.ac.at

CHAPTER OUTLINE

Abstract

Cells migrating in multicellular organisms steadily traverse complex three-dimensional (3D) environments. To decipher the underlying cell biology, current experimental setups either use simplified 2D, tissue-mimetic 3D (e.g., collagen matrices) or in vivo environments. While only in vivo experiments are truly physiological, they do not allow for precise manipulation of environmental parameters. 2D in vitro experiments do allow mechanical and chemical manipulations, but increasing evidence demonstrates substantial differences of migratory mechanisms in 2D and 3D.

Here, we describe simple, robust, and versatile "pillar forests" to investigate cell migration in complex but fully controllable 3D environments. Pillar forests are polydimethylsiloxane-based setups, in which two closely adjacent surfaces are interconnected by arrays of micrometer-sized pillars. Changing the pillar shape, size, height and the inter-pillar distance precisely manipulates microenvironmental parameters (e.g., pore sizes, micro-geometry, micro-topology), while being easily combined with chemotactic cues, surface coatings, diverse cell types and advanced imaging techniques. Thus, pillar forests combine the advantages of 2D cell migration assays with the precise definition of 3D environmental parameters.

1 INTRODUCTION

The movement of cells is essential for physiological and pathological processes such as development, tissue homeostasis, immune reactions, and tumor metastasis (Friedl & Weigelin, 2008; Kunwar, Siekhaus, & Lehmann, 2006; Reig, Pulgar, & Concha, 2014). Depending on the individual tissue and migratory path, cells face a variety of differently composed microenvironments that are crowded by cells and extracellular matrix (Charras & Sahai, 2014). For example, migratory immune cells encounter densely packed cellular microenvironments in lymph nodes, opposed to interstitial microenvironments dominated by extracellular matrix (Lämmermann & Germain, 2014; Worbs, Hammerschmidt, & Förster, 2017). Despite these tissue-specific differences, a hallmark of microenvironments in multicellular organisms is its three-dimensionality.

Whereas cell migration has traditionally been studied on two-dimensional (2D) surfaces leading to key discoveries of migratory mechanisms, increasing evidence demonstrates additional challenges and principles for motility in three-dimensional (3D) microenvironments (Friedl, Sahai, Weiss, Weiss, & Yamada, 2012; Lämmermann & Sixt, 2009; Leithner et al., 2016; Petrie & Yamada, 2016). For instance, while cellular migration on 2D strictly depends on cellular adhesions to the substrate, cells migrating in a confined manner between two surfaces can migrate independently of adhesions (such as in PDMS-based confiners or agarose assays) (Liu et al., 2015; Renkawitz et al., 2009). However, cells migrating in vivo are not only confined, but additionally encounter variable topographies, geometries and pore sizes (Stoitzner, Pfaller, Stössel, & Romani, 2002; Weigelin, Bakker, & Friedl, 2012; Wolf et al., 2009).

To faithfully mimic these 3D environmental parameters, we developed a "pillar forest" assay in which cells are not only confined between two surfaces, but also constantly face pillars on their migratory path (Fig. 1). Thereby, "pillar forests" represent tissue-mimetic flattened approximations of 3D matrices with precisely controlled environmental parameters. Modulation of pillar shapes, inter-pillar

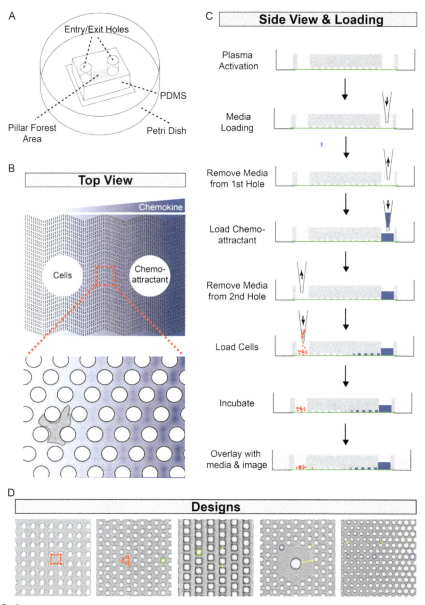

FIG. 1

Pillar forests to study cell migration in complex and controlled 3D environments. (A) Overview of a pillar forest device (with two entry/exit holes) attached to a Petri dish. (B) Top view onto the pillar forest area, showing the location of the manually punched holes directly into the pillar array. (C) Side view of the pillar forest device and loading procedure of cells and generation of a gradient of a chemoattractant (optional). (D) Examples of customizable geometrical parameters in pillar forests. Pillar forests can be designed with diverse geometries of pillar arrangements (red, examples of square and triangular lattice), pillar shapes (green, example of round and square pillars), as well as pillar sizes (violet circle) or pillar spacing (yellow lines, mimicking the pores size of the extracellular matrixes).

distances, and pillar arrangements enables the investigation of the individual contributions of topographies, geometries and pores sizes to cell migration.

We here describe the fabrication of "pillar forests" and their application to study cell migration in combination with chemotactic cues, surface coatings and advanced imaging techniques. We mainly focus on dendritic cells (DCs) as a model cell type for fast amoeboid cell migration, but would like to highlight our additional experience with other cellular examples such as other immune cell types (neutrophils, T cells, macrophages), cancer cells (HT1080), mesenchymal cells (fibroblasts) and fish keratocytes.

2 METHOD

2.1 OVERVIEW

Pillar forests are millimeter-sized devices (e.g., 3×9 mm) in which large arrays of micrometer-sized polydimethylsiloxane (PDMS)-based pillars are interconnecting two surfaces (Fig. 1). The relatively large size of the devices allows manual punching of two separate holes into the devices (e.g., by using a 2 mm-sized puncher). These holes enable washing of the device for equilibration with medium, surface coatings, loading of cells, and the establishment of a chemoattractant gradient (Fig. 1C). Upon loading, cells begin to migrate into the pillar forest between the entry holes. The architecture of the pillar forest can be customized by adjusting pillar shapes, pillar sizes, inter-pillar distances, and the geometry of pillar arrangements to the cell biological question of interest (Fig. 1D). We here describe every experimental step required to employ pillar forests as a controlled complex 3D environment, from the custom-made fabrication (Section 2.4) to coating, cell loading (Section 2.5), imaging and data analysis (Section 2.6).

2.2 MATERIALS

- Sylgard 184, 1 kg PDMS kit (Dowsil, USA)
- 4″ silicon wafer (Si-Mat, Germany)
- SU-8 GM 1040 (Gersteltec, Switzerland)
- SU-8 2005 (micro resist technologies, Germany)
- Propylene glycol monomethyl ether acetate (PGMEA), SU-8 developer (Merck, Germany)
- Trichloro(1H,1H,2H,2H-perfluorooctyl)silane, 97% (Merck, Germany)
- 70% Ethanol
- Silicone aquarium glue (e.g., Marina, Belgium)

2.3 EQUIPMENT

- 4″ diameter casting disk, 1 cm thick
- Harrick plasma cleaner, pdc-002 (Harrick Plasma, USA)

- Spin coater WS-650-23B (Laurell Technologies Corporation, USA)
- 200 g laboratory scale, 0.01 g resolution
- LINKCAD, Coreldraw X18 or Autocad software (http://www.linkcad.com, https://www.coreldraw.com, https://www.autodesk.com/products/autocad)
- Quartz 5″ photomask class 4, 1 μm resolution (JD Photo Data, UK)
- 2 mm Harris Unicore biopsy puncher (Merck, Germany)
- Hot plate with 1 °C resolution (e.g., digital hotplate SD160, Carl Roth, Germany)
- UV lamp, 365 nm collimated LED (M365L2-C1, Thorlabs, USA)
- EVG Mask Aligner 610 (EVG group, Austria)
- PL-360LP Photolithography mask aligner filter 215×215 mm (Omega Optical, USA)
- Sonicator (e.g., Elmasonic S30, Elma, Switzerland)
- Quadratic Petri dishes $120 \times 120 \times 17$ mm (Carl Roth, Germany)
- 60×15 mm tissue culture dishes with ø17 mm hole in the middle (home made, based on Corning Falcon easy grip cell culture dishes; commercially available dishes, e.g., from MatTek, USA might be also used)
- #1 or #1.5 cover slips
- *Optional*: Mixer-Defoamer, ARE-250 (Thinky, Japan)

2.4 FABRICATION OF PILLAR FORESTS

2.4.1 Photolithography and silanization of wafers

1. Design photomask in Coreldraw X18 or Autocad.
2. Convert file to Gerber format using LINKCAD.
3. Order Quartz 5″ photomask class 4, 1 μm resolution from JD Photo Data, UK.
4. Coat wafer with submicron base layer of SU8 GM1040, spin at 5000 rpm for 2 min in spin coater.
5. Pre-bake for 5 min at 95 °C.
6. Flood UV exposure with at least 100 mJ/cm^2.
7. Post-bake for 30 min at 95 °C.
8. Develop by submerging the wafer for 1 min in SU8 developer. Remove the developer with an airgun.
9. Spin coat SU8 2005 for 30 s to desired thickness. Use http://harnettlab.org/su8calc/ to calculate spin speed, e.g., 3.5 μm: 8390 rpm, 4 μm: 5000 rpm, 5 μm: 3000 rpm, 7 μm: 1370 rpm, 8 μm: 1000 rpm.
10. Soft bake for 2 min at 95 °C.
11. Expose wafer to 600 mJ/cm^2 through mask aligner filter. Exposure has to be optimized for your mask aligner of choice.
12. Post-exposure bake for 3 min at 95 °C.
13. Develop by submerging the wafer for 1 min in SU8 developer. Remove the developer with an airgun.
14. Hard-bake for 5 min at 135 °C.

15. Place wafers in a wafer carrier. Place inside vacuum desiccator inside a fume hood, together with an open 15 mL Falcon tube. Pipette 20 μL of silane into the tube and eject the tip in it.
16. Apply vacuum, seal the desiccator and turn off the vacuum pump.
17. Allow silane to vaporize for 1 h.
18. Vent the desiccator, screw the cap on the Falcon tube and dispose the tube as hazardous waste.
19. The wafers are now coated with a monolayer of hydrophobic silane from which PDMS can be peeled off easily.

2.4.2 Curing PDMS devices

1. Prepare a $6'' \times 6''$ square of aluminum foil.
2. Shape the aluminum foil around the wafer to make a circular tray. Place the wafer in the tray inside a square Petri dish.
3. Weigh 50 g of a 10:1 mixture of PDMS and PDMS curing agent in a Thinky mixer cup.
4. Mix in the Thinky mixer set to: 2000 rpm, 2 min; defoam: 2000 rpm, 2 min.
5. Pour the mix onto the wafer.
6. Place the wafer inside the square dish in a vacuum desiccator. Pump out the remaining air bubbles. Vent as necessary to pop air bubbles.
7. Cover the dish and place in an oven at 80 °C over night.

2.4.3 Dicing PDMS and plasma bonding

1. Cut off all PDMS and aluminum foil from the back of the wafer using a razor blade.
2. Peel off the PDMS carefully from the wafer. Work with its features facing upward and avoid touching the surface.
3. Cut PDMS into small pieces according to the design.
4. Punch holes for cell administration into the PDMS device with a 2 mm hole puncher. Alternatively a 12 gauge blunt sharpened needle can be used.
5. Clean glass cover slips by placing them into a beaker with isopropanol. Sonicate for 5 min in sweep mode. Replace isopropanol with pure ethanol and repeat sonication.
6. Remove ethanol and fill the beaker with distilled water. Take out slides with forceps and carefully dry them using an airgun. Store cleaned slides in a Petri dish.
7. Clean the PDMS device on all sides with scotch tape.
8. Place PDMS devices into beaker containing 70% ethanol.
9. Sonicate beaker for 5 min in sweep mode.
10. Setup hotplate at 85 °C.
11. Take PDMS devices out of the beaker with ethanol and dry them with an airgun.
12. Place PDMS devices with features facing upward together with cleaned glass slides into the plasma cleaner.

13. Plasma clean PDMS devices and glass slides at high power for 2 min.
14. Turn off the plasma cleaner and place glass slides on the hot plate. For bonding, carefully press PDMS devices, features facing downward onto the cover slips. Place onto hot plate for 1–2 h.
15. Glue cover slip bonded PDMS devices to Petri dishes with holes using aquarium glue. Let the glue harden in an oven at 80 °C over night.

2.4.4 Alternatives and troubleshooting

Troubleshooting: Photolithography and silanization of wafers—step 1. Consider the standard PDMS-based architectural design requirements such as the aspect ratio during the photomask design. Also, depending on wafer fabrication method, structures smaller than 1–2 μm may be difficult to generate. In our experience, it is highly convenient to design multiple different pillar forest designs onto the same wafer, which allows testing of different environmental parameters such as pillar distances suited for the cell type of interest.

Alternatives: Curing PDMS devices—step 3. Instead of using a Thinky mixer, PDMS and curing agent can be mixed with a scoop in a 50 mL Falcon tube. Large air bubbles can be removed by centrifugation (2000g, 20 min) before continuing with step 5.

Alternatives: Dicing PDMS and plasma bonding—step 5. Instead of using home made dishes with a hole, it is possible to use commercially available imaging dishes, e.g., from MatTek or FluoroDish. In this case the entire dish can be cleaned and plasma activated for PDMS bonding. If multiplexing is desired, 6- or 12-well plates with custom-made holes can be used to analyze multiple pillar forest devices in parallel.

Troubleshooting: Dicing PDMS and plasma bonding—step 14. In case of defective bonding make sure that the PDMS is dry and completely free of ethanol before plasma cleaning. If bonding keeps failing, ethanol cleaning (step 8) can be omitted.

With some devices pressing too hard might lead to collapse of the microstructures. On the other hand pressing too little might not lead to effective bonding. Therefore, the right amount of pressing has to be carefully determined for each device.

2.5 COATING, CHEMOKINE-GRADIENTS, AND CELL LOADING OF PILLAR FORESTS

2.5.1 Coating (optional) and pre-incubation of the device

Coating

1. Activate the chip in a plasma cleaner to generate a hydrophilic surface of the pillar maze and to ensure the proper entry of the solutions. In our experience,
 . 2 min on high mode has proven to be effective.

2. Pipette $\sim$10 μL of the coating solution into one of the wells. For example, we routinely use fibronectin (10 μg/mL) and PLL-PEG (500 μg/mL) diluted in the appropriate buffer.
3. Incubate 1 h at room temperature. Wash by removing the liquid from the holes and by flushing two times PBS through both holes.

Pre-incubation

4. Flush pre-warmed cell culture medium in the PDMS device. Incubate for 1 h in a humidified incubator at 37 °C; 5% CO_2 or in conditions suitable for your cell type.

Troubleshooting: Introduction of medium. Check that there are no air bubbles in your chips during either the coating or medium entry step. If the plasma cleaning is not efficient enough for the spreading of your coating solution, fill both entry holes and put the chip-containing dish under a vacuum for $\sim$2 min. Check under a microscope and repeat vacuum if necessary.

Pre-incubation of the device. PDMS is porous to gases, so equilibration in a proper medium and incubator at least 1 h prior experiment could be essential for both cell survival or the cellular processes of interest.

2.5.2 Generating a chemoattractant gradient (optional) and cell loading

1. Carefully remove $\sim$5 μL of the medium from one entry hole, and add $\sim$5 μL of chemokine. Avoid introducing air bubbles while doing so. For DCs we use CCL19 (2.5 μg/mL).
2. Pipette off most of the medium in the hole on the other side without generating air bubbles, and add $\sim$5 μL of cell suspension (50,000 cells).
3. Place the dish into the incubator for 1 h for DCs, then fill up the dish with medium so the chip is covered. In our experience, this will not disturb the gradient, but will eliminate differences in volumes in the two holes. Depending on the cell type, cells start to migrate in the pillar maze after some minutes (neutrophils, T cells, fish keratocytes), 0.5–3 h (dendritic cells), or >3 h (fibroblasts, macrophages).
4. Start the video-microscopy experiment.

Troubleshooting: Chemokine gradient generation. The distance between the chemokine- and cell hole should not be more than the diameter of the biopsy puncher (2 mm), and should stay constant between experiments. To monitor the gradient shape, one can add a fluorescent molecule with the same size as the chemokine (such as FITC-Dextran 10 kDa at 200 μg/mL) in the chemokine hole.

Introduction of cells. We recommend to resuspend your cells in cell culture media free of pH indicators (such as phenol red) to avoid background fluorescence. Cells may not enter the device when their concentration is too low. Adjust the concentration depending of your cell type; for leukocytes we use $5–10 \times 10^6$ cells/mL. Adjust the device geometry as well as its height depending on your cell type; for example, primary neutrophils or T cells may need a smaller confinement (2–3 μm) than dendritic cells (3–5 μm), T cell lines (4–6 μm), or fibroblasts (7 μm). Finally, cell entry can be delayed or impaired by inhibition of proteins (via gene silencing or drug treatment) that are essential for motility, and it should be accounted for with the proper controls during the experimental design and analysis.

2.6 IMAGING AND DATA ANALYSIS

Pillar forests enable high-resolution imaging experiments (such as TIRF and spinning disc microscopy), which are challenging in other 3D experimental setups (e.g., collagen, matrigel, in vivo). For example, in Fig. 2, we visualized filamentous actin (Lifeact-GFP, shown in red) in a T cell line migrating in a pillar maze (triangular lattice of 10 μm diameter pillars, separated by 5 μm, confinement height of 5 μm).

In general, standard image analysis software packages can be used to determine cell migration parameters in pillar forest assays. However, based on our experience, we found Fiji ("Fiji: an open-source platform for biological-image analysis") an its plugin TrackMate ("TrackMate: An open and extensible platform for single-particle tracking") especially helpful to track cells via their labeled nuclei to determine their average speed. Furthermore, we developed a matlab script that allows calculation of chemotactic indices from TrackMate files (Leithner et al., 2018). If cells are fluorescently labeled, the software package ilastik ("ilastik: Interactive Learning and Segmentation Toolkit") can be used to semi-automatically segment whole or parts of cells from the background. The resultant binary images can then be used to quantify parameters like protrusion dynamics as we have demonstrated recently (Leithner et al., 2016).

3 PITFALLS AND LIMITATIONS

While "pillar forests" mimic cellular migration mechanisms in 3D, physical differences to microenvironments composed of cells and/or extracellular matrix should be considered for the individual question of interest. This may include the stiffness difference of PDMS in comparison to cells/matrix, the size difference of pillars in comparison to fibers, and the fact that PDMS-based environments cannot be proteolyzed.

4 DISCUSSION

We here describe "pillar forests" as a reductionist approach to mimic complex 3D microenvironments with full control of environmental parameters such as topography and geometry. The flattened nature of the assay (cellular confinement in the z plane) allows precise analysis of cellular behaviors such as migration parameters, shape dynamics, and molecule localizations (Figs. 1 and 2). Moreover, "pillar forests" can be combined with gradients of chemoattractants, multi-well plates for

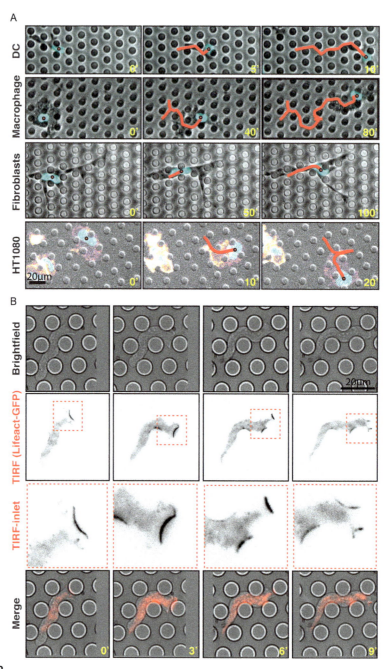

FIG. 2

Cellular examples migrating in pillar forests. (A) Exemplary dendritic cells (DCs), macrophages, fibroblasts and HT1080 cells migrating in pillar forests. Cyan depicts the nucleus (Hoechst). (B) TIRF microscopy of a Lifeact-GFP expressing T cell line (LMR7.5) migrating in a pillar forest.

screening approaches, coating or surface printing of molecules of interest (e.g., LAPAP (Schwarz et al., 2017) or UV patterning (Azioune, Storch, Bornens, Théry, & Piel, 2009)) and advanced live-cell imaging methods (e.g., TIRF and spinning disc microscopy) (Leithner et al., 2016).

Based on migration parameters such as velocity and cell shape, which faithfully mimic behavior in collagen matrices, we propose "pillar forests" as an alternative tissue-mimetic assay to collagen matrices, micro-channels (Vargas et al., 2016) and under agarose assays. Whereas collagen matrices are composed of bendable and digestible fibers closely mimicking the in vivo microenvironment, flat cellular confinement in under agarose assays enable spatio-temporal studies of fluorescently labeled molecules (e.g., by TIRF). "Pillar forests" combine the major advantages of both assays, while allowing the precise definition of environmental parameters (Fig. 3).

Our experience with DCs, T cells, neutrophils, macrophages, fibroblasts, HT1080 fibrosarcoma cells, and fish keratocytes suggests that pillar forests are broadly applicable to study migration of diverse cell types using different migratory strategies (e.g., amoeboid vs mesenchymal; single vs collective). Combined with the

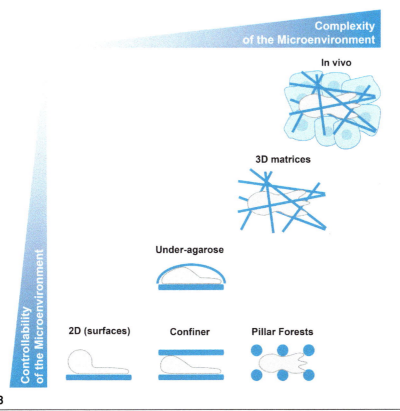

FIG. 3

Complexity and controllability of microenvironmental parameters in cell migration assays.

possibility to precisely manipulate the microenvironment, pillar forests enable the discovery of geometrical and topological principles of cell migration during immune responses, wound healing, and tumor dissemination. Further, we believe that pillar forests have the potential to decipher the principles of various other cell biological processes in tissue-mimetic complex 3D microenvironments, such as cell division (Lancaster et al., 2013; Spira et al., 2017), cell competition, cellular differentiation, and the cellular uptake of extracellular material (e.g., phagocytosis and macropinocytosis).

REFERENCES

Azioune, A., Storch, M., Bornens, M., Théry, M., & Piel, M. (2009). Simple and rapid process for single cell micro-patterning. *Lab on a Chip, 9*(11), 1640–1642.

Charras, G., & Sahai, E. (2014). Physical influences of the extracellular environment on cell migration. *Nature Reviews Molecular Cell Biology, 15*(12), 813–824.

Friedl, P., Sahai, E., Weiss, S., Weiss, S., & Yamada, K. M. (2012). New dimensions in cell migration. *Nature Reviews Molecular Cell Biology, 13*(11), 743–747. Nature Publishing Group.

Friedl, P., & Weigelin, B. (2008). Interstitial leukocyte migration and immune function. *Nature Immunology, 9*(9), 960–969.

Kunwar, P. S., Siekhaus, D. E., & Lehmann, R. (2006). In vivo migration: A germ cell perspective. *Annual Review of Cell and Developmental Biology, 22*, 237–265.

Lämmermann, T., & Germain, R. N. (2014). The multiple faces of leukocyte interstitial migration. *Seminars in Immunopathology, 36*(2), 227–251. Springer Berlin Heidelberg.

Lämmermann, T., & Sixt, M. (2009). Mechanical modes of "amoeboid" cell migration. *Current Opinion in Cell Biology, 21*(5), 636–644.

Lancaster, O. M., Le Berre, M., Dimitracopoulos, A., Bonazzi, D., Zlotek-Zlotkiewicz, E., Picone, R., et al. (2013). Mitotic rounding alters cell geometry to ensure efficient bipolar spindle formation. *Developmental Cell, 25*(3), 270–283.

Leithner, A., Eichner, A., Müller, J., Reversat, A., Brown, M., Schwarz, J., et al. (2016). Diversified actin protrusions promote environmental exploration but are dispensable for locomotion of leukocytes. *Nature Cell Biology, 18*(11), 1253–1259.

Leithner, A., Renkawitz, J., de Vries, I., Hauschild, R., Häcker, H., & Sixt, M. (2018). Fast and efficient genetic engineering of hematopoietic precursor cells for the study of dendritic cell migration. *European Journal of Immunology, 392*, 245. Wiley-Blackwell.

Liu, Y. J., Le Berre, M., Lautenschlaeger, F., Maiuri, P., Callan-Jones, A., Heuzé, M., et al. (2015). Confinement and low adhesion induce fast amoeboid migration of slow mesenchymal cells. *Cell, 160*(4), 659–672.

Petrie, R. J., & Yamada, K. M. (2016). Multiple mechanisms of 3D migration: The origins of plasticity. *Current Opinion in Cell Biology, 42*, 7–12.

Reig, G., Pulgar, E., & Concha, M. L. (2014). Cell migration: From tissue culture to embryos. *Development (Cambridge, England), 141*(10), 1999–2013. Oxford University Press for The Company of Biologists Limited.

Renkawitz, J., Schumann, K., Weber, M., Lämmermann, T., Pflicke, H., Piel, M., et al. (2009). Adaptive force transmission in amoeboid cell migration. *Nature Cell Biology, 11*(12), 1438–1443.

Schwarz, J., Bierbaum, V., Vaahtomeri, K., Hauschild, R., Brown, M., de Vries, I., et al. (2017). Dendritic cells interpret haptotactic chemokine gradients in a manner governed by signal-to-noise ratio and dependent on GRK6. *Current Biology: CB, 27*, 1314–1325. Elsevier.

Spira, F., Cuylen-Haering, S., Mehta, S., Samwer, M., Reversat, A., Verma, A., et al. (2017). Cytokinesis in vertebrate cells initiates by contraction of an equatorial actomyosin network composed of randomly oriented filaments. *eLife, 6*, 983.

Stoitzner, P., Pfaller, K., Stössel, H., & Romani, N. (2002). A close-up view of migrating Langerhans cells in the skin. *The Journal of Investigative Dermatology, 118*(1), 117–125.

Vargas, P., Chabaud, M., Thiam, H.-R., Lankar, D., Piel, M., & Lennon-Duménil, A.-M. (2016). Study of dendritic cell migration using micro-fabrication. *Journal of Immunological Methods, 432*, 30–34.

Weigelin, B., Bakker, G.-J., & Friedl, P. (2012). Intravital third harmonic generation microscopy of collective melanoma cell invasion. *Intravital, 1*(1), 32–43.

Wolf, K., Alexander, S., Alexander, S., Schacht, V., Schacht, V., Coussens, L. M., et al. (2009). Collagen-based cell migration models in vitro and in vivo. *Seminars in Cell & Developmental Biology, 20*(8), 931–941.

Worbs, T., Hammerschmidt, S. I., & Förster, R. (2017). Dendritic cell migration in health and disease. *Nature Reviews Immunology, 17*(1), 30–48.

Measuring spontaneous neutrophil motility signatures from a drop of blood using microfluidics

Sinan Muldur, Anika L. Marand, Felix Ellett[1], Daniel Irimia[1]

BioMEMS Resource Center, Department of Surgery, Massachusetts General Hospital,
Harvard Medical School, Shriners Burns Hospital, Boston, MA, United States
[1]Corresponding authors: e-mail address: fellett@mgh.harvard.edu; dirimia@mgh.harvard.edu

CHAPTER OUTLINE

Abstract

Neutrophils play an essential role in the protection against infection, as they are the most numerous circulating white blood cell population and the first responders to injury. Their numbers in blood are frequently measured in the clinic and used as an indicator of ongoing infections. During inflammation and sepsis, the ability of neutrophils to migrate is disrupted,

Methods in Cell Biology, Volume 147, ISSN 0091-679X, https://doi.org/10.1016/bs.mcb.2018.07.005

which may increase the risk of infection, even when the neutrophil count is normal. However, measurements of neutrophil migration in patients are rarely performed because of the challenges of performing the migration assays in a clinical setting. Here, we describe a microfluidic assay that measures the spontaneous neutrophil migration signatures associated with sepsis. The assay uses one droplet of patient's blood in a microfluidic device, which circumvents the need for neutrophil isolation from blood. This assay may also be useful for the study of the effect of various immune modulators on neutrophil migration behavior from healthy volunteers and patients.

1 INTRODUCTION

The study of innate immune responses in patients, particularly those of neutrophils, is under increasing investigation. Neutrophils constitute the major circulating white blood cell population and are the first responders to tissue injury. Neutrophil count is part of the standard blood analysis in clinical laboratories. Neutropenia is an abnormally low neutrophil count. It can be inherited or acquired and increases susceptibility to bacterial or fungal infections and impairs the resolution of these infections. Neutrophilia, on the other hand, is an increased number of neutrophils, most often caused by ongoing inflammatory and infectious diseases. In some cases, such as burn injuries (Lavrentieva et al., 2007) the neutrophil count alone is not an accurate indicator of immune status. Recently, neutrophil functional competence has been proposed as more valuable in a broad range of conditions, such as sepsis.

Sepsis is a life-threatening organ dysfunction caused by a dysregulated host response to infection (Singer et al., 2016). If not diagnosed early and managed promptly, sepsis can lead to shock, multiple organ failure, and death. Despite advances in healthcare, existing epidemiologic studies suggest that sepsis remains a considerable burden which affects more than 30 million people worldwide every year (Fleischmann et al., 2016) and is increasing at a rate of 1.5% annually (Angus & Van der Poll, 2013). In addition to its high mortality rate, this global crisis places a significant clinical and economic burden on the healthcare system. Sepsis is the leading cause of hospitalization in the US costing more than $23 billion each year (Torio & Moore, 2016).

Early diagnosis of sepsis is critical, as mortality is decreased by 7.6% per hour when antibiotics are administered early to patients (Kumar et al., 2006). However, early diagnosis remains a challenge as the pathophysiology of sepsis is complex and remains incompletely understood. There are several biomarkers of sepsis, including C-reactive protein (CRP), procalcitonin (PCT), interleukin-6 and reactive oxygen species but none proved to be reliable and clinically effective (reviewed by Liu et al., 2016). A major obstacle to accurate diagnosis of sepsis is the differentiation between sepsis and systemic inflammatory response syndrome (SIRS).

In the context of sepsis, neutrophil dysfunction likely contributes to both weak immune responses and off-target organ damage (Brown et al., 2006). Neutrophils from septic patients lose the ability to respond appropriately to chemotactic signals (Jones et al., 2014) and have altered antimicrobial activity (Solomkin, 1990).

Neutrophil migration has been monitored in several studies as a potential indicator of inflammatory status or infections (Hoang et al., 2013). However, standard migration assays such as Boyden chambers (Boyden, 1962), Dunn and Zigmond chambers (Zicha et al., 1991; Zigmond, 1977) and micropipette techniques (Gerisch & Keller, 1981) have significant limitations in quantifying the dynamic nature of the migration process. Moreover, these assays are not adequate for a clinical setting as they are time-consuming and require large volumes of blood. The development of microfluidic-based assays compatible with direct patient blood analysis is addressing many of these limitations and is increasing the accuracy of the measurements. A recent study measured the CD64 expression by neutrophils in blood from 450 patients, using a microfluidic device that enabled the prediction of patient sepsis prognosis (Hassan et al., 2017). In another study, a microfluidic assay using a P-selectin-coated surface was developed to purify neutrophils from whole blood and study chemotaxis (Sackmann et al., 2012). Moreover, the direct measurement of neutrophil motility in a droplet of blood, in the presence of other blood cells and serum factors, without the requirement for any neutrophil purification, has also been described (Ellett et al., 2018). Direct characterization of neutrophil migration using microfluidic devices can be applied to multiple donor species, and a recent study has revealed significant differences among migration counts, velocity, and directionality among neutrophils from mice, rats, and humans (Jones et al., 2016).

Our group has previously developed a microfluidic device that identified a sepsis-specific spontaneous migration signature displayed by isolated neutrophils originating from septic patients, which enabled the prediction of septic patients with 80% sensitivity and 77% specificity (Jones et al., 2014). Inspired by these results, our group has recently engineered a microfluidic device that measures, from one droplet of diluted blood, the spontaneous migration behaviors of neutrophils in the context of sepsis (Ellett et al., 2018). From 42 patients, the use of whole blood increased the performance of the assay to 97% sensitivity and 98% specificity for sepsis.

In this chapter, we describe in detail the fabrication steps of this microfluidic device, the experimental and analytical procedures used to investigate septic patient blood samples. We also employ the device to test the effect of various immune modulators on the spontaneous neutrophil migration behavior from healthy volunteers.

2 MICROFLUIDIC DEVICE

Here we describe the fabrication and preparation of a microfluidic assay that enables measurement of spontaneous neutrophil motility from a diluted blood sample. This method can be used to investigate and diagnose patient blood such as septic patients (Ellett et al., 2018) or study the effect of various immune modulators on spontaneous neutrophil migration behavior by spiking whole blood from healthy volunteers.

This protocol is dedicated to the measurement of spontaneous neutrophil migration only, as no chemotactic gradient is generated within these devices prior sample loading.

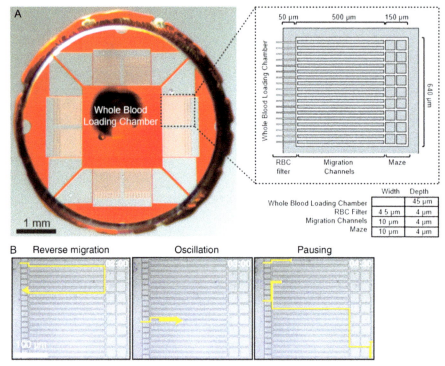

FIG. 1

A microfluidic device to assay spontaneous neutrophil motility. (A) Macroscopic image of the microfluidic device indicating the loading chamber and one of the eight migration mazes. Right, a magnified view (dashed box) showing a detailed diagram of the neutrophil migration maze. (B) Still images extracted from a time-lapse video obtained from a sample from a patient with sepsis, showing examples of behavior identification from neutrophil tracks. The time stamps is in h:min.

From Ellett, F., Jorgensen, J., Marand, A. L., Liu, Y. M., Martinez, M. M., Sein, V., et al. (2018). Diagnosis of sepsis from a drop of blood by measurement of spontaneous neutrophil motility in a microfluidic assay. Nature Biomedical Engineering, 2(4), 207.

Each microfluidic device is composed of eight migration mazes, two on each side of a square whole blood-loading chamber. Each neutrophil migration maze is composed of a red blood cell (RBC) filter, a series of migration channels and a maze (Fig. 1).

2.1 MICROFLUIDIC DEVICE FABRICATION

1. Design a device using AutoCAD and print chrome masks for photolithography.
2. Fabricate the master mold on a silicon wafer, in a clean room, using standard photolithographic techniques. Spin-coat the silicon wafer, with a first

4-μm thick epoxy-based negative photoresist layer (SU-8, Microchem, Newton, MA) to define the migration channels and a second 45-μm epoxy layer to define the whole blood-loading chamber.

3. Pattern the wafer by sequential ultraviolet light exposure through two photolithographic masks and process according to the manufacturer's instructions.

4. Use the patterned wafer as a mold to cast polydimethylsiloxane (PDMS, Sylgard 184, Ellsworth Adhesives, Wilmington, MA) device. Mix vigorously PDMS (20 g) with curing agent (2 g) and pour carefully onto mold.

5. Use a vacuum desiccator for at least 1 h to degas PDMS.

6. Bake and cure PDMS microfluidic device for at least 12 h in an oven set to 75 °C.

7. Cut the PDMS from the master mold.

8. Punch the delivery port using a 1.2-mm punch (Harris Uni-core, Ted Pella).

9. Punch the entire device out of the PDMS using a 5-mm puncher (Harris Uni-core, Ted Pella).

10. Remove particles from device surfaces using adhesive tape.

11. Oxygen plasma treat a 35-mm glass-bottom multiwell plate (P06G-1.5-20-F, MatTeK Co., Ashland, MA) or a single small glass-bottom dish (P35G-0-20-C, MatTeK Co., Ashland, MA) twice; once alone for 35 s and then again, along with the devices (face up) for another 35 s.

12. Using tweezers, bond devices by carefully placing them face down on the glass-bottom multiwell plate or Petri dish. Apply slight pressure on the device to evacuate any air trapped between the device and the glass.

13. Bake plate with bonded devices on a hot plate set to 75 °C for 10 min.

Note: Devices bonded on glass-bottom multiwell plates (e.g., 6-well plates) can be used in a confocal microscope and standard inverted fluorescent microscope. The devices bonded on glass-bottom small Petri dishes are used for the CytoSMART imaging system (Fig. 2).

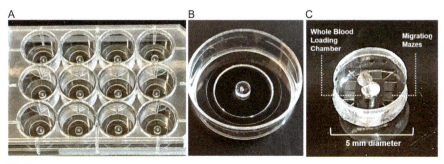

FIG. 2

Bonded microfluidic device. (A) Device bonded to 35-mm glass-bottom 12-well plate (MatTeK Co., Ashland, MA). (B) Device bonded to a single small 35-mm glass-bottom Petri dish (MatTeK Co., Ashland, MA). (C) Magnified view of microfluidic device composed of a loading chamber in the center and two mazes on each side (eight mazes in total per device).

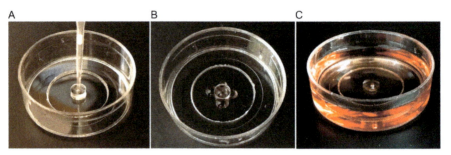

FIG. 3

Microfluidic assay preparation. (A, B) Priming device with IMDM and 20% FBS by pipetting in, on top and around the device. (C) Microfluidic device submerged in media.

2.2 MICROFLUIDIC ASSAY PREPARATION

1. Prime devices with 50 μL Iscove's modified Dulbecco's medium (IMDM, ThermoFisher Scientific) containing 20% fetal bovine serum (FBS, Life Technologies) by pipetting in, on top, and around the device (Fig. 3).
2. Vacuum devices for 10 min and allow to equilibrate for another 10 min until all channels are filled. By applying a vacuum to the device, the solution is instilled into the channels of the device as gases in the channels diffuse into the de-gassed PDMS.
3. Confirm wetting of device channels under microscope. No bubbles should be present within the device.
4. Fill the well containing the device with fresh IMDM containing 20% FBS until the top of the device is completely submerged under liquid.

Note: These devices can be prepared and stored at 4 °C up to 1 month before use.

3 SAMPLE PREPARATION AND LOADING

Here we describe the possibility of performing two different types of experiments using human blood samples. First, by acquiring a patient blood sample, this assay can enable in a short amount of time early-diagnosis of septic condition (Ellett et al., 2018). Second, by collecting peripheral or capillary blood from a healthy donor, this assay can be employed to study the effect of immune modulators on the induction of neutrophil motility and thus attempt to recapitulate sepsis-like neutrophil phenotype by spiking blood.

3.1 HUMAN NEUTROPHILS FROM PATIENT PERIPHERAL BLOOD

1. Collect or order peripheral blood sample from patients in heparin-coated vacuum tubes (Vacutainer, Becton Dickinson), prefereably from indwelling lines.

Use blood sample within 1 h after collection. (*Patient samples must be obtained after written informed consent and through procedures approved by an Institutional Review Board.*)

2. Prepare 50 μL IMDM with 20% FBS stained with Hoechst 33342 dye (ThermoFisher) at 32 μM (Staining is necessary for fluorescent microscopy).
3. Add 50 μL whole blood sample to 50 μL stained media (1:1 dilution).
4. Mix by gently pipetting once.
5. Incubate the blood and Hoechst stain for 5–10 min at room temperature to allow for fluorescent staining of cell nuclei.
6. Extremely gently, pipette 1.5 μL of stained blood into the loading chamber using a gel-loading tip (Eppendorf), taking care to draw the tip out of the device while dispensing.

3.2 HUMAN NEUTROPHILS FROM HEALTHY PERIPHERAL BLOOD

1. For spiking experiments, collect or order peripheral blood from healthy volunteers aged 18 years or older in heparin-coated vacuum tubes and use as soon as possible, within 1 h after collection. A sample volume of 0.01–1 mL of blood is usually sufficient.
2. Prepare the immune modulators solution in IMDM with 20% FBS at twice the target modulator concentration stained with Hoechst (32 μM).
3. Add 50 μL whole blood sample to 50 μL immune modulator in stained media (1:1 dilution).
4. Mix by gently pipetting once.
5. Incubate spiked blood sample for at least 15 min at room temperature.
6. Load the blood sample as previously described.

Note: Hoechst stain is necessary for only for fluorescent microscopy.

3.3 HUMAN NEUTROPHILS FROM HEALTHY CAPILLARY BLOOD

1. For spiking experiments, prepare 1 mL of IMDM with 20% FBS in a 2 mL Sodium Heparin BD Vacutainer. Mix vigorously.
2. Prepare the immune modulator solution in previous IMDM with 20% FBS (with Sodium Heparin) at twice the target concentration stained with Hoechst (32 μM).
3. From a consenting volunteer aged 18 years or older, disinfect one finger using an alcohol swab and let dry. Promptly prick side of finger with a 1.5 mm × 30G contact-activated lancet (BD Microtainer). Wipe away the first drop of blood with a sterile gauze pad.
4. Coax blood out of pricked site by gently pressing around the area. Using a pipette tip, collect blood and pipette into a 1.5 mL Eppendorf tube.
5. Mix blood 1:1 with previously immune modulators solution.
6. ~20–50 μL is usually the maximum amount of blood collected from a finger prick. The volumes used for the 1:1 dilution step should be prepared accordingly.
7. Mix by gently pipetting once.

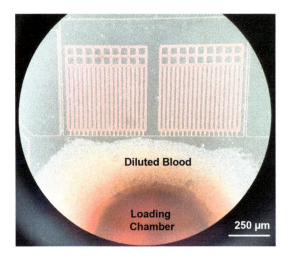

FIG. 4

Microscopic view of a loaded microfluidic chip. 10× Objective (Olympus CKX41). The scale bar is 250 μm.

8. Incubate spiked blood sample for at least 15 min at room temperature.
9. Load the sample as previously described.

Note: Use Hoechst stain only if using fluorescent microscopy. For controls, use the Heparin containing media without immune modulator.

Tips
○ Loading too much blood or pressing the pipette too strongly may lead to RBCs entering the mazes, causing blockage and preventing neutrophils from migrating into the side channels. At the end of loading, there should be a uniform circle of whole blood in the center of the device, not yet in contact with the side channels and mazes.
○ Load devices when near microscope to avoid shaking dishes during transportation.
○ Do NOT overload the wells with media, as this will cause media to spill out the sides when the lid is placed on (Fig. 4).
○ Make sure devices are fully covered by media prior to loading. Under-filling the well will cause the sample to flow through and out the device.

4 MICROSCOPY
4.1 BRIGHTFIELD MICROSCOPY—CytoSMART LIVE-CELL IMAGING
The CytoSMART LUX2 system (Eindhoven, The Netherlands) is a compact and incubator-proof inverted microscope for brightfield live-cell imaging. It is composed of a 10× fixed objective, 5MP CMOS Sensor, and has a 2.4 × 1.5 mm field of view.

1. Connect CytoSMART microscope to a PC tablet and place device in an incubator or warm room at 37 °C.
2. Open CytoSMART software and choose between "zoom in" and "zoom out" option in the menu tab depending on how many migration mazes is desired in the field of view. *The "zoom out" option allows two mazes to be imaged but may make it harder to see the migrating neutrophils. Using the "zoom in" option allows only one maze in the field of view, but the higher magnification enables easier tracking of the moving neutrophils.*
3. Align device (glass-bottom Petri dish) by manually adjusting position and make sure microscope is focused on the maze(s) using the "focus bar" on the bottom of the screen.
4. Press "start experiment" and enter experiment name.
5. Choose snapshot interval for 10 s (this setting may require a software updated from CytoSMART if not already implemented).
6. Hit "start" and the microscope will automatically acquire images for 4 h.

Note: The field of view of the CytoSMART system allows one to collect data from a maximum of two mazes out of eight mazes present in one device. Determination of the final "Sepsis score" requires multiplying the score obtained from two mazes by a factor 4 to reflect the whole device (Fig. 5).

4.2 TIME-LAPSE FLUORESCENT MICROSCOPY

1. Alternatively, one could record the spontaneous neutrophil migration into mazes with a fully automated time-lapse fluorescent Nikon TiE inverted wide-field microscope (10 × or higher) with a biochamber heated to 37 °C with 5% CO_2 (or equivalent microscope).

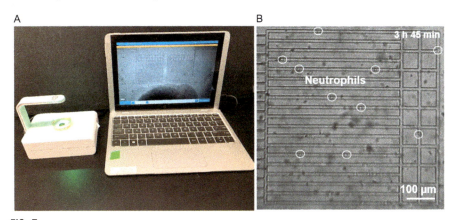

FIG. 5

Brightfield microscopy—CytoSMART live-cell imaging. (A) CytoSMART mini-microscope connected to tablet, imaging a microfluidic device. (B) Microscopic brightfield image obtained from CytoSMART showing spontaneously migrating neutrophils (white circles). The scale bar is 100 μm.

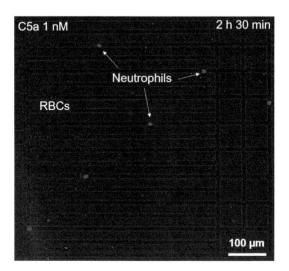

FIG. 6

Spontaneous neutrophil migration observed after spiking healthy blood with immune modulator. Fluorescent microscopic image showing nucleus of migrating neutrophils stained with Hoechst (32 μM). Several red blood cells (RBCs) can be observed inside the migration channels. The scale bar is 100 μm.

2. *Important*: Place the multiwell plate onto microscope stage to set up and save points BEFORE loading any diluted blood samples into devices. Each microfluidic device provides eight fields of views, each containing one migration maze. Each field is imaged every 2 min to enable accurate tracking of cell motility, for 4 h.
3. Remove the multiwell plate and load the blood sample into each device, as previously described.
4. Place the multiwell plate back onto microscope stage and make sure the plate is in the exact same location as when setting points. This can be checked by looking at the live view of the maze at the first x/y point. Start experiment as soon as possible, ideally within minutes after loading the sample.

Note: Images of the loading chamber can be recorded as well and employed to approximate the total number of neutrophils in the device and to calculate the percentage of neutrophils migrating (Fig. 6).

5 DATA ANALYSIS

Neutrophil spontaneous migration can be tracked using ImageJ/Fiji analysis software, from either brightfield images (acquired using a CytoSMART microscope) or fluorescent time-lapse images (acquired using a Nikon TiE microscope). Neutrophil tracking is key for quantifying the various spontaneous migration parameters necessary for the calculation of the sepsis score.

The experiments conducted with CytoSMART are saved in a local folder containing a compilation of each snapshot in a JPG format, which is automatically saved under the Browse tab under "Menu." The experiments using a fluorescent Nikon TiE and the Nikon NIS-Elements software are saved in ND2 format in a local folder. Both formats are compatible with ImageJ/Fiji analysis software.

5.1 BRIGHTFIELD MANUAL TRACKING

Image pre-processing
1. Import "image sequences" using ImageJ/Fiji. For the CytoSMART JPG files, select "virtual stack" when prompted.
2. Crop images to get one migration maze and rotate images such that whole blood chamber is on the left edge of the field of view (optional).
3. Adjust brightness and contrast.
4. Subtract background by selecting "Process/Subtract background" commands. Decrease rolling ball radius to 20 pixels to increase visibility of cells (optional).

Manual tracking
1. Open "tracking" then "manual tracking" in the "Plugins" menu.
2. Press "show parameters," fill in the appropriate "time interval" and "x/y calibration" information. *Important*: For the CytoSMART using low magnification ("zoom out"), the calibration is 1.945 μm/pixel. If using other microscopes the calibration will be different. To perform the calibration, use the known length of the maze (650 μm) and divide by the measured length of the maze in pixels.
3. Press "add track." On the image sequence, scroll through time until observing a neutrophil passing through the RBC filter and entering the main channel. Press on the neutrophil. A new "result" window will appear showing track number, distance, velocity, and other parameters. Every time a new point is pressed, the following image in the sequence appears. Follow neutrophil migration until the end of the experiment while taking note of possible migration phenotypes (arrest, oscillations, and retrotaxis). When finished, press "end track" and follow another neutrophil by pressing "add track," which will start back at the beginning of the sequence of stacks.
4. After tracking every neutrophil entering the maze, select, copy and export tracking results in Excel (MS Office) or equivalent software. The Results window can also be directly saved in an XLS format which can open in Excel.
5. Track neutrophils in all mazes of the same device/condition.
6. Tracking in instances of numerous neutrophil migration to ensure complete tracking of all cells:
   i. Once the first track is completed select "overlay lines" or "overlay dots" under "drawing." This will produce a new stack with a traced track of the cell as lines or dots, depending on which option is selected.
   ii. Each time a new overlay stack is produced, the previous overlay stack may be deleted. Check to make sure that the last track that was overlaid is on the new overlay stack before deleting the previous.

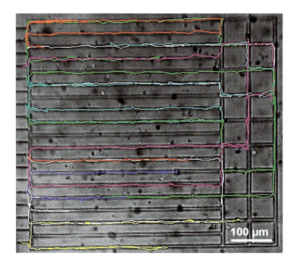

FIG. 7

Manual tracking of migrating neutrophils using ImageJ/Fiji. Microscopic images were obtained using the CytoSMART mini-microscope. Tracks of different colors represent distinct neutrophils. The scale bar is 100 μm.

    iii. Select "add track" with the new overlay stack and track another cell.

    iv. A new "result" window will pop up with the previous track(s) and the current track.

    v. *Important:* Make sure to select "overlay lines" or "overlay dots" after completing each cell's track. Always keep the most recent "results" window with the latest number of tracks.

    vi. After the last cell is overlaid, the stack sequence will show all the tracked cells on one stack, which can be saved and used to count any of the above described phenotypes used to calculate the sepsis score (Fig. 7).

Note: Whenever an error is made when tracking a cell, use the "delete last point" function in the tracking menu. This will remove the last point clicked. To delete an entire track: select track number (which is found under the "delete last point" tab) and select "delete track."

Sepsis score calculation

1. Sum the individual cell's traveled distances in both mazes and find the average distance traveled of all neutrophils in that device. Divide this by four.

2. Determine the number of migrated neutrophils from the results window. From the final overlain stacks of each maze determine the total number of phenotypes observed in both mazes of the same device.

3. Calculate the score for two mazes:

Sepsis score $= N*(O+P+R+AD)/10^3$

where N, is the total number of migrated neutrophils,

O, the total number of oscillation phenotype, defined as "the total number of cells that switch direction twice in a channel and migrate for more than 15 μm in each segment,"

P, the total number of arrest phenotype, defined as "a cell with zero velocity,"

R, the total number of retrotaxis phenotype, defined as "any cell that leaves the maze back into the central chamber,"

AD, average distance of all neutrophils traveled in the two mazes, divided by four.

4. As each device is composed of eight mazes in total, to calculate the sepsis score (Ellett et al., 2018), the score obtained above is multiplied by 4.
5. A blood sample with a sepsis score over 30 is considered septic (Fig. 7).

5.2 FLUORESCENT AUTOMATED TRACKING—TRACKMATE (IMAGEJ/FIJI)

Image pre-processing
1. On ImageJ/Fiji, open ND2 file (or another image sequence).
2. In "BioFormat Import Options," make sure to open the file as a "Hyperstack" and that "Specify range for each series" is the only box checked in the memory management setting. In color options, select "composite" color mode.
3. Select Series (each of them is one x/y point): select one series to analyze.
4. In "BioFormats Range Option," use default settings.
5. Rotate image such that the blood-loading chamber is on the left (optional).
6. Crop the images selecting the area containing migration channels and maze only (exclude the RBC filter channels).
7. Adjust brightness and contrast for both channels.
8. Select the Hoechst/Blue channel and scroll to a time point in which migrating cells are visible.

Tracking using TrackMate
1. In "Plugins," open "tracking" then "TrackMate."
2. Select "Log detector."
3. Select the "channel segment number" corresponding to Hoescht/Blue channel (2, if using brightfield and Hoescht channels only). Estimate Blob Diameter: 10–15 μm with a threshold of 5 μm. Preview to verify accuracy. If identifying object other than cells, increase the threshold until not recognized.
4. Initial Thresholding: select all.
5. Select a view: "Hyperstack Displayer."
6. Select filter on spots: "Uniform Color."
7. Select a Tracker: "LAP tracker."
8. Set the tracking parameters. Select the appropriate time interval. For images acquired at 2 min time interval, select the "linking distance" at 150–250 μm, the "gap-closing" distance is 50–100 μm and the "gap-closing max frame gap" at 2 μm. Select "feature penalties" for "Quality" and "X" to a value of 1 for both linking distances and gap-closing. This prevents horizontal tracking between

two channels. Note: Y penalties should be used instead of X penalties if the direction of the main channels runs horizontal.

9. Scroll through time and verify that the track matches cell migration trajectory. If gaps are visible, return to "Settings for tracker" to increase "Linking distance" until gap disappears. If tracker joins tracks that are from two different cells, decrease "linking distance."

10. Set filter on tracking: "Track ID." Additional filters can be added to enhance tracks such as "Track displacement" which eliminates small tracks that are not from neutrophils.

11. Display Options: select "Analysis." Three windows open with different tracking data.

12. Select, save and export set of data displaying distances for each track (cell) present in the maze.

13. Close all windows and start from the beginning with the next series.

14. Sepsis score or neutrophil migration phenotypes calculations can then be performed as described previously.

6 CONCLUSION

Neutrophils are an important component of the innate immune system. Obtaining a deeper understanding of neutrophil biology in the context of homeostasis, infection, and inflammation will provide a valuable resource to both clinical and research communities. Microfluidic devices provide an opportunity to study neutrophil biology in the context of small volumes of minimally processed whole blood, enabling studies in the presence of other blood cells and plasma, and greatly reducing the costs traditionally associated with isolation. Here, we provide detailed protocols for the use of recently-described microfluidic devices, for studying spontaneous neutrophil motility in the context of disease, and stimulation by candidate immune modulators.

ACKNOWLEDGMENTS

This project was supported by funding from the National Institutes of Health, National Institute of General Medical Sciences (Grant GM092804) and Shriners Hospitals for Children. Microfluidic devices were manufactured at the BioMEMS Resource Center at Massachusetts General Hospital, supported by a grant from the National Institute of Biomedical Imaging and Bioengineering (Grant EB002503).

REFERENCES

Angus, D. C., & Van der Poll, T. (2013). Severe sepsis and septic shock. *New England Journal of Medicine, 369*(9), 840–851.

Boyden, S. (1962). The chemotactic effect of mixtures of antibody and antigen on polymorphonuclear leucocytes. *Journal of Experimental Medicine, 115*(3), 453–466.

Brown, K. A., Brain, S. D., Pearson, J. D., Edgeworth, J. D., Lewis, S. M., & Treacher, D. F. (2006). Neutrophils in development of multiple organ failure in sepsis. *The Lancet, 368*(9530), 157–169.

Ellett, F., Jorgensen, J., Marand, A. L., Liu, Y. M., Martinez, M. M., Sein, V., et al. (2018). Diagnosis of sepsis from a drop of blood by measurement of spontaneous neutrophil motility in a microfluidic assay. *Nature Biomedical Engineering, 2*(4), 207.

Fleischmann, C., Scherag, A., Adhikari, N. K., Hartog, C. S., Tsaganos, T., Schlattmann, P., et al. (2016). Assessment of global incidence and mortality of hospital-treated sepsis. Current estimates and limitations. *American Journal of Respiratory and Critical Care Medicine, 193*(3), 259–272.

Gerisch, G., & Keller, H. U. (1981). Chemotactic reorientation of granulocytes stimulated with micropipettes containing fMet-Leu-Phe. *Journal of Cell Science, 52*(1), 1–10.

Hassan, U., Ghonge, T., Reddy, B., Jr., Patel, M., Rappleye, M., Taneja, I., et al. (2017). A point-of-care microfluidic biochip for quantification of CD64 expression from whole blood for sepsis stratification. *Nature Communications, 8*, 15949.

Hoang, A. N., Jones, C. N., Dimisko, L., Hamza, B., Martel, J., Kojic, N., et al. (2013). Measuring neutrophil speed and directionality during chemotaxis, directly from a droplet of whole blood. *Technology, 1*(1), 49–57.

Jones, C. N., Hoang, A. N., Martel, J. M., Dimisko, L., Mikkola, A., Inoue, Y., et al. (2016). Microfluidic assay for precise measurements of mouse, rat, and human neutrophils chemotaxis in whole blood droplets. *Journal of Leukocyte Biology, 100*(1), 241–247.

Jones, C. N., Moore, M., Dimisko, L., Alexander, A., Ibrahim, A., Hassell, B. A., et al. (2014). Spontaneous neutrophil migration patterns during sepsis after major burns. *PLoS One, 9*(12), e114509.

Kumar, A., Roberts, D., Wood, K. E., Light, B., Parrillo, J. E., Sharma, S., et al. (2006). Duration of hypotension before initiation of effective antimicrobial therapy is the critical determinant of survival in human septic shock. *Critical Care Medicine, 34*(6), 1589–1596.

Lavrentieva, A., Kontakiotis, T., Lazaridis, L., Tsotsolis, N., Koumis, J., Kyriazis, G., et al. (2007). Inflammatory markers in patients with severe burn injury: What is the best indicator of sepsis? *Burns, 33*(2), 189–194.

Liu, Y., Hou, J. H., Li, Q., Chen, K. J., Wang, S. N., & Wang, J. M. (2016). Biomarkers for diagnosis of sepsis in patients with systemic inflammatory response syndrome: A systematic review and meta-analysis. *SpringerPlus, 5*(1), 2091.

Sackmann, E. K., Berthier, E., Young, E. W., Shelef, M. A., Wernimont, S. A., Huttenlocher, A., et al. (2012). Microfluidic kit-on-a-lid: A versatile platform for neutrophil chemotaxis assays. *Blood, 120*(14), e45–e53.

Singer, M., Deutschman, C. S., Seymour, C. W., Shankar-Hari, M., Annane, D., Bauer, M., et al. (2016). The third international consensus definitions for sepsis and septic shock (sepsis-3). *JAMA, 315*(8), 801–810.

Solomkin, J. S. (1990). Neutrophil disorders in burn injury: Complement, cytokines, and organ injury. *The Journal of Trauma, 30*(12 Suppl), S80–S85.

Torio, C., & Moore, B. (2016). *National inpatient hospital costs: The most expensive conditions by payer.* https://www.hcup-us.ahrq.gov/reports/statbriefs/sb204-Most-Expensive-Hospital-Conditions.jsp. Accessed 5 June 2018.

Zicha, D., Dunn, G. A., & Brown, A. F. (1991). A new direct-viewing chemotaxis chamber. *Journal of Cell Science, 99*(4), 769–775.

Zigmond, S. H. (1977). Ability of polymorphonuclear leukocytes to orient in gradients of chemotactic factors. *The Journal of Cell Biology, 75*(2), 606–616.

Directing cell migration on flat substrates and in confinement with microfabrication and microfluidics

Emilie Le Maout*,†,‡,§,¶, **Simon Lo Vecchio***,†,‡,§,¶, **Alka Bhat***,†,‡,§,¶
Daniel Riveline*,†,‡,§,¶,1

**Laboratory of Cell Physics ISIS/IGBMC, CNRS and University of Strasbourg, Strasbourg, France*
†*Institut de Génétique et de Biologie Moléculaire et Cellulaire, Illkirch, France*
‡*Centre National de la Recherche Scientifique, UMR7104, Illkirch, France*
§*Institut National de la Santé et de la Recherche Médicale, U964, Illkirch, France*
¶*Université de Strasbourg, Illkirch, France*
1*Corresponding author: e-mail address: riveline@unistra.fr*

CHAPTER OUTLINE

Methods in Cell Biology, Volume 147, ISSN 0091-679X, https://doi.org/10.1016/bs.mcb.2018.06.006

Abstract

Cell motility has been mainly characterized *in vitro* through the motion of cells on 2D flat Petri dishes, and in Boyden chambers with the passage of cells through sub-cellular sized cavities. These experimental conditions have contributed to understand important features, but these artificial designs can prevent elucidation of mechanisms involved in guiding cell migration *in vivo*. In this context, microfabrication and microfluidics have provided unprecedented tools to design new assays with local controls in two and three dimensions. Single cells are surrounded by specific environments at a scale where cellular organelles like the nucleus, the cortex, and protrusions can be probed locally in time and in space. Here, we report methods to direct cell motion with emphasis on micro-contact printing for 2D cell migration, and ratchetaxis/chemotaxis in 3D confinements. While sharing similarities, both environments generate distinct experimental issues and questions with potential relevance for *in vivo* situations.

1 INTRODUCTION

Cell motility plays a key role during development (Helvert, Storm, & Friedl, 2018). When its normal mode of operation is altered, defects appear in tissues, and they can be involved in diseases such as cancer progression. Many genes and pathways have been shown to be important for regulating cell motions *in vivo*, and their roles were further characterized *in vitro*.

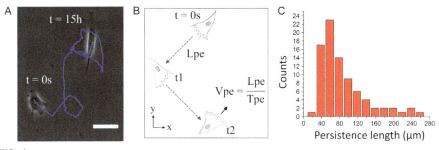

FIG. 1

Characterization of NIH3T3 fibroblast migration on a 2D flat substrate. (A) Typical trajectory of a cell moving randomly on a flat coverslip, the blue trace shows the trajectory. Scale bar 100 µm. (B) Schematics showing the migration of a cell, Lpe represents the persistence length, Tpe the persistence time and Vpe the persistence speed; pausing time and number of turns per unit time can also be measured. (C) Distribution of a parameter, here the persistence length, for NIH3T3 fibroblasts migrating on flat surface ($n=84$ trajectories).

Indeed cell motion can be recapitulated on 2D flat coverslips or on Petri dishes: a polarized cell migrates in one direction, pauses, potentially changes polarity and migrates in another direction. Their typical motions over days are represented in Fig. 1. Five parameters can encode cell motility. Persistence length is the length traveled "straight" by the cell during a certain time with no pause. This time is called persistence time. Persistence speed is deduced from the ratio of these two values (see Fig. 1B). Next, cell pauses and this phase can be characterized by its duration time; the number of turn per unit time captures the changes in directions. Measurements of these five parameters can lead to interesting results to characterize cell motion in a generic way (see Fig. 1C; Caballero, Voituriez, & Riveline, 2014).

1.1 PETRI DISHES AND BOYDEN CHAMBERS: THEIR LIMITS

Experiments have been mainly performed so far in conditions which can be viewed as "artificial" compared to *in vivo* conditions. For example, cells are cultured and observed in flat Petri dishes (see Figs. 1A and 2A), and this method was kept probably for historical reasons (Petri, 1887). In this setup, cells have random shapes and this can alter their migration; also, they evolve on 2D flat substrates which constitutes an important difference with 3D physiological environments. At the stage of plating, cells can appear as single cells or as groups of cells with an ill-controlled number of cells per group. These varying initial conditions can lead to bias in the study of motility. Other assays using Boyden chambers, for example, can also generate artificial conditions (see Fig. 2C; Boyden, 1962): cells are placed in an upper chamber and are exposed to a membrane with cavities of controlled dimensions, typically 10 µm in diameter and 10 µm in depth. In the presence of chemical gradients between the top and bottom compartments, cells eventually migrate. However, the motion itself

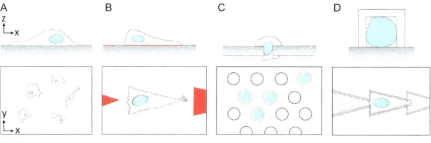

FIG. 2

Cell motility on different configurations. (A) Cell on Petri dish/flat 2D substrate. (B) Cell on 2D ratchet. (C) Cell on a "membrane" in a Boyden chamber experiment. (D) Cell in 3D confined ratchet. The nucleus is depicted in blue. In all panels, we represent x–z plane (top) and x–y plane (bottom).

is restricted to a small fraction of the cell size. This 10 μm cavity limits considerably the potential extrapolation to *in vivo* situations. As a result, with Petri dishes and Boyden chambers, the study of cell migration lacks control and the assays can be far from reproducing physiological conditions.

1.2 THE MICROFABRICATION REVOLUTION AND ITS IMPACTS

The usage of techniques developed through microfabrication and microfluidics (Whitesides, 2006) have opened a new era for the study of cell migration (Caballero et al., 2014; Cramer, 2010; Hawkins et al., 2009; Jiang, Bruzewicz, Wong, Piel, & Whitesides, 2005; Prentice-Mott et al., 2013). When studying 2D cell migration, the shape of cells can be imposed on surfaces through adhesive micro-contact printed proteins surrounded by cell repellent (see Fig. 2B; Théry & Piel, 2009). Cell symmetry and its effect on migration can be analyzed quantitatively and encoded in physical models (Fig. 5; Caballero et al., 2014; Hawkins et al., 2009). In addition, cells can be confined in 3D through the appropriate preparation of micro-channels (see Fig. 2D); this can reproduce conditions where single cells are trapped in blood vessels with diameters smaller than cell dimensions. This configuration can also lead to controlled cell symmetries (see Figs. 7 and 8). In both cases, on flat 2D printed substrates and in 3D confined geometries, motions can be acquired in the presence of chemical gradients (Comelles et al., 2014; Prentice-Mott et al., 2013).

These developments have been possible because polydimethylsiloxane (PDMS)—the main material to prepare these assays—is biocompatible and easy to manipulate. The material is still liquid right after addition of the cross-linker and can be deposited on molds and cured in a 65 °C oven. The PDMS solid mold can be replicated with negative/positive combinations with up to nanometer resolutions in the reproduction of motifs (Qin, Xia, & Whitesides, 2010). Altogether, cell shapes and dimensions are now controlled with microfabrication, and these methods have shown to be essential for obtaining new standardized tests.

We report below experimental procedures to prepare these assays for the same cell type, i.e. NIH 3T3 fibroblasts. Any cell can be used, but the fact of probing the same cell line allows to identify changes in behaviors between experimental conditions set by the assay, such as speeds and directions.

1.3 DESIGNING THE MOTIFS: SCALING ARGUMENTS

In order to match the motility assays to each cell type, cell dimensions need to be determined first. The mean area of resting cells on flat coverslips allows to evaluate the optimal surface area needed to micro-contact print adhesive motifs on 2D flat surfaces. Cells will spread on these motifs with their spontaneous resting shape (Fig. 2B). In addition, measure of the mean cell diameter after trypsinization and before cells start spreading, helps to evaluate the cell volume and this can guide the design of motifs for 3D confinements (Fig. 2D). As reported above, the cell symmetry can also be modified: cells can adopt disk shapes (2D; Caballero et al., 2014) or move along straight channels (3D; Prentice-Mott et al., 2013), and their symmetry can be broken while keeping the projected area/volume constant in 2D/3D assays respectively. In fact, several studies have reported that the presence of asymmetric patterns in the environment can direct cell motion. This kind of motion has been named *ratchetaxis* (Caballero, Comelles, Piel, Voituriez, & Riveline, 2015). In 2D configuration, lines of periodic triangles can be generated to simplify the ratchetaxis motion along a single direction. For the 3D case, lines of periodic connected triangles can be generated to study the motion in confined environments with a local broken symmetry. In addition, the junction/gap between subsequent motifs can be controlled. In 2D, a cell on a motif will have different dynamics depending on the gap between neighboring motifs. In 3D confined situations, the opening width of connecting triangles will test different regimes in cell motility.

Altogether, cell area in 2D and cell volume in 3D will set rules for the design of the microfabricated unit. Gaps/junctions between motifs will test cell probing and confinement respectively. The sum of microfabricated unit length and junction's length can be integrated in a lattice unit or in cell dimension to compare between conditions and between cell types (L is about 100 μm for NIH 3T3 cells, see Fig. 1A). Cells trajectory can be then calculated as lattice unit/cell length dimensions. Symmetry/asymmetry will be generated by keeping constant area/volume in 2D/3D respectively.

Finally, the cell velocity should be taken into account for the design of experiments. Specifically, trajectories should be sufficiently long to acquire reliable statistics; this can be evaluated with cell velocity. For example, let us say that cells move at 10 μm/h speed with a typical cell dimension of 100 μm. If we want to follow the cell migration for 10 times the length of a cell (around 1000 μm), taking into account the frequent pauses on 2D flat surfaces (see Fig. 1A), this would require an acquisition of at least 24 h. In turn, the number of periodic lattice units should be prepared accordingly to allow cells to potentially move along this 1000 μm long trajectory.

With these simple orders of magnitude in mind, the masks containing the motifs can be designed for microfabrication.

2 MICROFABRICATION OF MOTIFS

2.1 DESIGNING THE MOTIFS

Design of the mask can be done using different softwares, e.g. Clewin (Freeware) or Autocad®. A light sensitive material called *photoresist* is used during microfabrication: it is spin-coated on a wafer before exposure of the wafer to UV light when substrate and mask are in close contact (Fig. 3).

Motifs are designed according to the type of photoresist, positive or negative. Here in our study, epoxy based photoresist SU-8 (MicroChem) was used. SU-8 being a "negative" photoresist becomes insoluble to the SU-8 developer when exposed to UV light. The unexposed regions will be stripped off during the development process leaving behind the micro-structures. Masks have to be prepared accordingly, in order

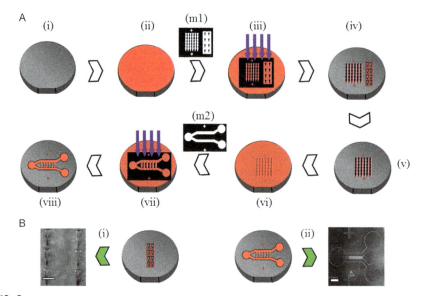

FIG. 3

Microfabrication for the study of 2D and 3D ratchetaxis. (A) Steps followed to obtain motifs. (i) Silicon wafer (ii) spin-coated with SU-8 (2005, 2025) according to the required height/depth. (iii) UV exposure of the spin-coated wafer through the desired mask (m1) for the first layer. (iv) Developed motifs of the first layer with 5 μm high connected triangles; left side of the wafer and 25 μm deep first layer of separated triangles; right side of the wafer. (v) Wafer with 5 μm high connected triangles (vi) spin-coated with photoresist; SU-8 2025 for 50 μm thickness. (vii) UV exposure of the spin-coated wafer through the desired mask (m2) for the second layer. (viii) Developed motifs of the second layer; connected triangles with the reservoir. (B) Developed resin motifs on a Si wafer of (i) 25 μm thick first layer of separated triangles (right side: holes on wafer; left side: image of the wafer, scale bar 100 μm); and (ii) 5 μm thick connected triangles with 50 μm thick reservoir (left side: pillars on wafer: right side, image of the wafer, scale bar 1 mm).

to fabricate either holes or pillars on the wafer. It is important to have dark motifs ("+" polarity) on the mask, so that the UV light cannot pass through, in turn fabricating holes on the wafer. Similarly, for fabricating pillars, transparent motifs ("−" polarity) have to be designed, which would allow the UV light to pass through the motifs, cross-linking the exposed photoresist. Structures fabricated in such a way have excellent thermal and mechanical stability.

After the masks with the desired dimensions are printed, motifs are prepared through microfabrication (see Section 2.2.1.3).

2.2 MICROFABRICATION

2.2.1 Protocol

2.2.1.1 Materials

- Silicon wafers (Si-Mat, cat. PRIME/76.5 mm/381 mm)
- Wafer tweezers
- Chromium/plastic photomask (Selba) with printed motifs to micro-fabricate
- SU-8 photoresist (MicroChem cat. SU8/2025/0.5)
- SU-8 developer (Chimie Tech Services, cat. DevSU8/4)
- Acetone
- 2-Propanol
- Ethanol 70%

2.2.1.2 Equipment

- Mask aligner (SUSS MicroTec, cat. MJB3)
- Spin-coater (Laurell Technologies, cat. WS-400B-6NPP)
- Hot plates (65 °C and 95 °C)
- Disposable graduated dropper (to remove air bubbles)

2.2.1.3 Method

Two sets of motifs are required for the processes reported in Sections 3–5. Section 3: micro-contact printing requires one layer of the photoresist (25 µm). The depth is tuned to have an optimal aspect ratio: if the pillars are too high and the motif area is too small, pillars will bend during stamping and the quality of the printed patterns will not be optimum. Section 4: open microchannel configuration requires one layer of the photoresist (25 µm): height of the photoresist pillars is tuned to have an optimal aspect ratio, and PDMS should be high enough to trap cells (see Fig. 7A). Sections 4 and 5: closed microchannel configuration requires two layers; a first layer of 5 µm thickness and a second layer of 50 µm thickness. The first layer is the microchannel which confines the cells from the top, whereas the second layer constitutes the reservoir for cells and should be high enough to allow cell movement and positioning.

The following steps are performed in a clean room, if available (see Fig. 3):

1. Clean the silicon wafers with first acetone and then ethanol. Dry the wafer using a nitrogen stream after each solvent cleaning. Solvent cleaning insures

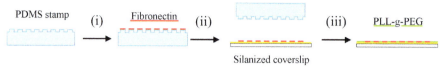

FIG. 4

Micro-contact printing procedure: (i) Incubation of the PDMS stamp with fibronectin. (ii) Stamping. (iii) Passivation with PLL-g-PEG.

complete removal of oils and organic residues from the surface of silicon wafers; while they do remove contaminants, solvents themselves actually leave residues on the surface of wafers as well. For this reason, a two-solvents method is implemented to ensure that the wafer is contaminant-free. After this step, heat the wafers at 200°C for 15–30 min for a complete removal of humidity from the wafer surface.

2. Spin-coat the first photoresist layer according to the target thickness. To obtain separated triangular patterns for micro-contact printing (see Fig. 4) and open micro-channels (see Fig. 7), pour the photoresist (SU-8 2025) on the wafer and remove air bubbles with the aid of a plastic dropper. Set a two-step spinning process on the spin-coater. Spin at 500 rpm for 10 s with an acceleration of 100 rpm/s during the first step to homogenize the layer over the wafer. Next, spin the wafer containing the resin at 3000 rpm for 30 s with acceleration of 300 rpm/s for the second step. This should result in a photoresist layer of 25 μm thickness (see Fig. 3A (ii)).

3. To obtain the first layer for the closed connected triangular microchannel (see Fig. 8), a 5 μm thick layer is required. After repeating step (2) using the desired photoresist (SU-8 2005), spin at 3000 rpm for 30 s with acceleration of 300 rpm/s for the second spin step. This should result in a photoresist thickness of 5 μm (see Fig. 3A (ii)).

4. After obtaining the evenly spread photoresist, pre-bake has to be done to ensure a firm attachment of the photoresist to the wafer. For a 25 μm thick layer, this step has to be done during 5 min at 95°C and for a 5 μm thick layer during 2 min at 95°C.

5. After pre-bake, make a firm contact between the wafer and the mask containing the respective motifs. A defective contact will lead to ill-defined motifs (see Troubleshooting: Tr.1; Fig. 10A). The masks should have negative polarity for 25 μm thick layer (Separated Triangles) to give holes and positive polarity for 5 μm thick layer (Connected Triangles) to obtain pillars on photoresist (SU-8) layer. After tight contact with the mask (see Fig. 3A (m1)) cross-link the photoresist by exposing the wafer to UV irradiation with a dose of $150 \, \text{mJ/cm}^2$ for 25 μm thick Separated Triangles and $105 \, \text{mJ/cm}^2$ for 5 μm thick Connected Triangles (see Fig. 3A (iii)). The time of exposure will depend on the power/wavelength of the device (e.g., mask aligner) for the UV light exposure.

6. As soon as the exposure is over, for Separated Triangles: follow a post-bake of 1 min at 65 °C and then 5 min at 95 °C; and for Connected Triangle: 3 min at 95 °C. An image of the mask should be visible on the (SU-8) photoresist coating.

7. After post-bake, develop the structures by immersing them in the SU-8 developer solution, while gently agitating the container; 4 min for 25 μm thick Separated Triangles and 1 min for 5 μm thick Connected Triangles layer. This would strip off the non-cross-linked resin, leaving behind the required motifs (see Fig. 3A (iv)). Finally, rinse the surface with 2-propanol to remove the leftover photoresist.

8. Dry the wafer containing motifs, using a nitrogen stream after rinsing with 2-propanol.

9. After preparing a 5 μm thick layer of Connected Triangles (see Fig. 3A (v)), a second layer has to be prepared in order to obtain a microchannel of connected triangles with a reservoir for cells on either side of the Connected Triangles, in order to introduce cells in microchannels (see Fig. 8B).

10. The second layer (for reservoir) is recommended to be 50 μm thick to facilitate the entry of cells with the cell culture media. For this purpose, follow step (2) and spin the wafer containing the resin (SU-8 2025) at 1700 rpm for 30 s with an acceleration of 300 rpm/s for the second spin step (see Fig. 3A (vi)).

11. Then follow a pre-bake of 6 min at 95 °C.

12. Before the next step of UV exposure for cross-linking, check on the silicon wafer that the mask for the second layer containing motif for reservoir (see Fig. 3A (m2)) is perfectly aligned with the developed first layer of Connected Triangles. For alignment purpose, it is recommended to design masks with two crosses each, on both mask 1 (containing microchannel motif) and mask 2 (containing the reservoir motif) exactly at the same position. During microfabrication and alignment process, both crosses should superimpose between the first photoresist (SU-8) layer and the second mask to ensure optimal alignment of the first and second layers (see Troubleshooting: Tr. 2)

13. Once the alignment is performed, secure the tight contact between the wafer and the mask (for reservoir). Cross-link the photoresist by exposing the wafer to UV irradiation with a dose of 150–155 mJ/cm^2 for a 50 μm layer of reservoir (see Fig. 3A (vii)).

14. As soon as exposure is over, follow a post-bake of 1 min at 65 °C and then 6 min at 95 °C. After post-bake at 95 °C, an image of the mask should be visible on the photoresist coating.

15. Next, develop the structures by immersing the sample in the SU-8 developer solution while gently agitating the container for 5 min. This would strip off the non-cross-linked resin, leaving behind motifs (see Fig. 3A (viii)). Finally, rinse the surface with 2-propanol to remove the leftovers of photoresist.

16. After the silicon wafers are ready with motifs (with one or two layers), do a "hard bake" step at 150 °C for a couple of minutes. This is useful for annealing any surface cracks that may have appeared after development; this step is relevant to all layer thicknesses.

17. After obtaining the motifs for Separated Triangles (25 µm thick layer) and Connected Triangles (first layer: 5 µm, second layer: 50 µm), cover the wafers with PDMS (cross-linker:prepolymer [1:10]).

18. Remove all air bubbles through desiccation, and finally cure overnight at 65 °C on a leveled surface.

19. Once cured, peel off the PDMS, the chip is ready.

3 RECTIFYING CELL MOTION WITH ASYMMETRICAL PATTERNS: MICRO-CONTACT PRINTING OF ADHESIVE MOTIFS

3.1 THE DESIGN OF PATTERN

The scaling arguments are presented in Caballero et al. (2014) for this assay.

3.2 MICRO-CONTACT PRINTING

3.2.1 Protocol

In this section, we detail the protocol for printing adhesive motifs on a glass coverslip. We use fluorescent fibronectin to visualize motifs with microscopy (other proteins could also be used). Poly-L-lysine-g-polyethylene glycol (PLL-g-PEG) is used for surface passivation (see Fig. 4).

3.2.1.1 Materials
- "Piranha" solution (sulfuric acid: hydrogen peroxide [7:3])
- Glass coverslips, N°1–25 mm diameter
- 3-(Mercapto)propyltrimethoxysilane
- Phosphate-buffered saline 1 × (PBS)
- 10 µg/mL Rhodamine labeled Fibronectin (FN) (Cytoskeleton, cat. FNR01-A) in PBS
- 0.1 mg/mL PLL-g-PEG diluted in HEPES 10 mM (SuSoS AG, cat. SZ33-15)
- PDMS (Dow Corning, cat. DC184-1.1)
- Milli-Q Water
- Ethanol 70%
- Parafilm

3.2.1.2 Equipment
- Plasma Cleaner (Diener Electronic, cat. ZeptoB)
- Oven
- Sonicator
- Vacuum pump
- Desiccator

3.2.1.3 Methods
1. Clean the coverslips inside "Piranha" solution during 10 min (this step should be performed cautiously under a hood). Bubbles will form, check that coverslips stay immersed throughout the process. Carefully rinse the glass coverslips

with Milli-Q water. Then, sonicate one by one each coverslip in beakers during 5 min and dry them with a nitrogen stream. Finally, put them in an oven at 65 °C for 10 min.

2. Place the "Piranha" cleaned coverslips inside the desiccator with a small Petri dish filled with 100 μL of 3-(mercapto)propyltrimethoxysilane. Generate a vacuum, then close the pump and let the silane deposit on substrates for 1 h.

3. Place the silanized coverslips for at least 90 min at 65 °C (up to 4 h); note that non-bound silane is very sensitive to temperature and humidity.

4. Meanwhile, cut the PDMS to have 1 cm × 1 cm stamps; sonicate them for 5 min in ethanol and dry them with a nitrogen stream.

5. Activate the surface of each stamp with oxygen plasma (air can also be used). This step will make sure that the surface is hydrophilic and suitable for protein incubation and binding. Be sure that patterns face up during activation.

6. Incubate activated stamps with a 100 μL drop of 10 μg/mL solution of Rhodamine labeled FN for 1 h.

7. After incubation, remove the drop with a pipette and quickly dry the stamp with a nitrogen stream. It should take only few seconds. You can also let the stamp dry at room temperature for about 5 min. This step is critical and differences in drying will impact the overall quality of patterns (see Troubleshooting: Tr.3 and Tr.4).

8. Put the stamp (face with the patterns down) on the silanized coverslip with a 50 g weight on top of it. Wait for 30 min and gently remove weight and stamp from the coverslip. Store the patterned glass coverslips in PBS at 4 °C. At this point, patterns can be stored up to one week, even if immediate usage is recommended.

9. Deposit a 100 μL drop of PLL-g-PEG on a piece of parafilm and put the patterned face of the coverslip onto the drop. Incubate for 20 min. This step will passivate the surface and will decrease adhesion of cells outside the patterns.

3.2.2 2D cell migration experiment

The cells should be carefully placed on the micro-contact printed motifs. Proper washing allows to generate samples with cells exclusively on motifs. Cells should be plated at low density. This will help prevent them to migrate along the same path and in turn potentially "collide." Such phenomenon would interfere with cells trajectories. Finally, low serum condition allows to keep standard cell motility while preventing cell division during migration.

3.2.2.1 Materials
- NIH3T3 cells (or other migratory fibroblasts)
- Dulbecco's Modified Eagle Medium (DMEM) 4.5 g/L glucose 1% Penicillin–streptomycin
- Bovine Calf Serum (BCS)
- Leibovitz's (L-15) medium
- Trypsin–EDTA
- Cell counter

- Pipettes
- 5 mL Petri dishes (60 mm in diameter)

3.2.2.2 Equipment
- Epifluorescence microscope with phase contrast and temperature control
- Sterile hood
- Incubator with CO_2
- Centrifuge

3.2.2.3 Methods
At this stage, all steps should be performed under a sterile hood with a laminar flow.

1. Trypsinize a 5 mL Petri dish with adherent cells 3 min, and add 3 mL of DMEM 10% BCS.
2. Centrifuge cells at 500 rpm during 3 min and re-suspend the pellet in 5 mL of DMEM 10% BCS.
3. Count cells and place 2000 cells on the micro-patterned coverslip.
4. Incubate cells for 30 min at 37 °C with 5% CO_2. Wash out the sample to remove non-adherent cells, and replace medium by L-15 with 1% BCS. This low serum condition will ensure standard cell motility while reducing cell division during the 48 h experiments.
5. Go to the microscope and image patterns and cells (see Troubleshooting: Tr.5) (Fig. 5).

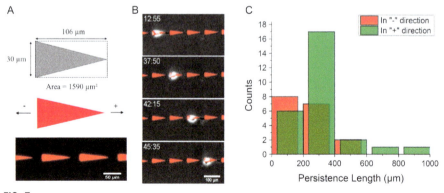

FIG. 5

(A) Pattern characteristics and dimensions (up) and motifs printed on a surface with micro-contact printing (bottom). Patterns are stamped with rhodamine labeled fibronectin. For clarity, tip direction is set as "+" and triangle base direction as "−" (middle). (B) Sequence of a NIH3T3 cell moving on triangles separated by 22 μm gaps. $T=0$ corresponds to the beginning of acquisition (shortly after plating). Migration is directed toward the "+" direction in this sequence; Scale bar 50 μm. Time in hh:mm. (C) Persistence length distribution of NIH3T3 cells migrating in "+" direction and in "−" direction. Motion is directed, or *rectified*, to the + direction on average.

4 CONTROL OF CELL MIGRATION IN CONFINED 3D ENVIRONMENT

4.1 THE DESIGN OF PATTERN

Along the scaling arguments (see Section 1), we measured cell volumes after trypsinization. We obtained a mean diameter for individual rounded cells of $D_{cell} = 14.4 \pm 1.6\,\mu m$ ($n = 388$ cells, s.d.), which corresponds to a mean volume of $V_{cell} = 1560 \pm 520\,\mu m^3$. Nuclei were stained on spread cells using Hoechst, and their volume and height were measured. The mean volume was $V_{nucleus} = 740 \pm 180\,\mu m^3$ ($n = 199$ nuclei, s.d.) and the mean height $H_{nucleus} = 8.5 \pm 1.4\,\mu m$ ($n = 199$ nuclei, s.d.). Altogether, we used V_{cell}, $V_{nucleus}$, $H_{nucleus}$ to design the 3D triangle channel motif: each triangular motif has the volume of a NIH3T3 fibroblast ($1500\,\mu m^3$). Taking into account these values, we selected the height of the microchannel of $5\,\mu m$ to allow confinement of cell and nucleus. We calculated the xy area for each triangle to be $300\,\mu m^2$. After selecting an angle of $16°$ similar to the angle used in triangles on 2D ratchetaxis and to further confine the nucleus, we selected an opening width of $4\,\mu m$ between connected ratchets. Using the following parameters (area: $300\,\mu m^2$, angle: $16°$ and opening width: $4\,\mu m$), base and length of a ratchet were calculated. They are equal respectively to $14\,\mu m$ and $34\,\mu m$ (see Figs. 7 and 8). In ratchetaxis experiments, we used this motif in a closed microchannel configuration, where cells were confined from the top and sides (see Fig. 8); and in open microchannel configuration, cells were confined only from the sides (see Fig. 7).

In order to have a large number of trajectories per experiment, we designed a mask with 50 rows of connected triangles. For the closed configuration, two layers of SU-8 are required. The first layer contains connected ratchets of $5\,\mu m$ height. The second layer allows to fabricate the cells reservoir of $50\,\mu m$ height. In order to generate chemical gradients, the second layer has a Y-shaped channel of $50\,\mu m$ height (see Figs. 3 and 9B). For the open microchannel configuration, a layer of $25\,\mu m$ height SU-8 is required in order to trap cells (see Fig. 7A).

4.2 CONFINED OPEN MICROCHANNEL CONFIGURATION

4.2.1 Protocol

4.2.1.1 Materials

- Masks/silicon wafer
- PDMS
- Glass coverslips (N°1–25 mm diameter)
- Chlorotrimethylsilane (Sigma, cat. 92360)
- 2 mL Petri dish (35 mm in diameter)

4.2.1.2 Equipment

- Plasma cleaner
- Tweezer
- Blade

- Oven
- Spin-coater
- Vacuum pump
- Desiccator

4.2.1.3 Methods

1. Pour PDMS (cross-linker:prepolymer [1:10]) on a silicon wafer. Remove air bubbles by placing the wafer into desiccator for 1 h. Cure at 65 °C for at least 4 h.
2. Cut PDMS with a blade and remove the PDMS 1 stamp from the wafer (see Fig. 6A (i)).
3. Activate the surface of PDMS stamp 1 with oxygen plasma. Put it inside the desiccator with a small Petri dish filled with 100 μL of chlorotrimethylsilane. Generate a vacuum, then close the pump and let the silane deposit on the substrate for 30 min.
4. Pour PDMS (cross-linker:prepolymer (1:10)) on PDMS 1 placed on Petri dish. Check that patterns are "face up." Remove air bubbles by placing the wafer into the desiccator for 1 h. Cure at 65 °C for at least 4 h (see Fig. 6A (ii)).
5. Cut PDMS with a blade and remove the PDMS stamp 2 from the Petri dish containing PDMS stamp 1.
6. Activate with oxygen plasma the surface of PDMS stamp 2. Put it inside the desiccator with a small Petri dish filled with 100 μL of chlorotrimethylsilane. Generate a vacuum, then close the pump and let the silane deposit on the substrate for 30 min.
7. Spin liquid PDMS on PDMS stamp 2. Set a two-step spinning process on the spin-coater. Spin at 500 rpm for 10 s during the first step to homogenize the PDMS layer over the PDMS stamp 2. The speed used in the second step will

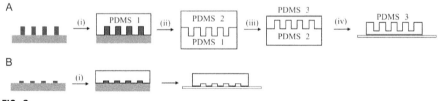

FIG. 6

Schematics of the method to obtain closed (A) and open (B) microchannel configuration. (i) PDMS is poured on SU-8 wafers and cured 4 h at 65°C. In closed configuration, microchannels are 5 μm height. (ii) After activation of PDMS and coverslip with oxygen plasma, they are both bonded together. In open configuration, microchannels are 25 μm height. (ii) PDMS 1 is cut, activate with oxygen, treated with chlorotrimethylsilane and PDMS is poured on it and cured 4 h at 65 °C. (iii) PDMS 2 is cut, activated with oxygen, treated with chlorotrimethylsilane and PDMS is spin-coated on top of PDMS 2 at 2000 rpm to obtain 40 μm PDMS layer. (iv) After curing at 65 °C for 4 h and activation with oxygen plasma, PDMS 3 is bonded to a coverslip and incubated overnight at 65 °C to consolidate the bonding. PDMS 2 is then peeled off.

define the height of PDMS stamp 3. Spin-coating at a speed of 2000 rpm allows the spreading of PDMS at a height of approximately 40 μm. Cure at 65 °C for at least 4 h (see Fig. 6A (iii)).

8. Activate both coverslip and PDMS stamp 3 (see Fig. 6A (iv); blue side) with oxygen plasma. Bind the blue surface side of PDMS stamp 3 to the coverslip. Incubate them at 65 °C overnight to secure the binding between the PDMS stamp 3 and the glass coverslip.

9. With a tweezer, gently peel off the PDMS stamp 2 from the PDMS stamp 3 (which is attach to the coverslip).

4.2.2 Cell migration experiment
4.2.2.1 Materials
- NIH3T3 cells (or other migratory fibroblasts)
- DMEM 4.5 g/L glucose 1% Penicillin–streptomycin
- BCS
- L-15 medium
- Trypsin–EDTA
- Cell counter
- Pipettes
- 5 mL Petri dishes (60 mm in diameter)

4.2.2.2 Equipment
- Epifluorescence microscope with phase contrast and temperature control
- Sterile hood
- Incubator with CO_2
- Centrifuge

4.2.2.3 Methods
At this stage, all steps should be performed under a sterile hood with a laminar flow. We report now how to place cells in the sample.

1. Before the experiment, activate the surface of the stamp with oxygen plasma. This step will make the surface hydrophilic and will facilitate media distribution within the sample. Place the open microchannel sample under UV for 10 min sterilization.

2. Trypsinize a 5 mL Petri dish with adherent cells 3 min and add 4 mL of DMEM 10% BCS. Count cells.

3. Centrifuge cells at 500 rpm during 3 min and re-suspend cells in DMEM 10% BCS at a density of 100,000 cells/mL.

4. Place 100 μL of the cell suspension on the open microchannel configuration.

5. Incubate 30 min at 37 °C with 5% CO_2. Wash out the sample to remove non-adherent cells and replace the medium with L-15 containing 1% BCS.

6. Go to the microscope and acquire cell migration with a 10 min interval.

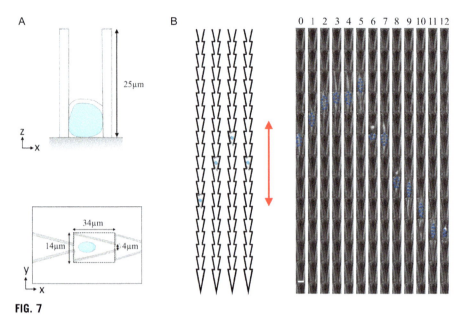

FIG. 7

(A) Dimension of the triangular 3D open configuration, *xz* plane (top) and *xy* plane (bottom). (B) Left: Schematic of the experimental setup. Cells are seeded in the middle of open ratchet microchannel. Right: Time-lapse images of a NIH3T3 fibroblast migrating in an open ratchet microchannel configuration. Blue dots outline the nucleus over time. Scale bar 10 μm, time in hour.

In open ratchet microchannel configuration, cells seeded in channels migrate longer and preferentially in the "+" direction after probing in both directions (see a typical example Fig. 7B (manuscript in preparation)).

4.3 CONFINED CLOSED MICROCHANNEL CONFIGURATION

4.3.1 Protocol

4.3.1.1 Materials

- Masks/silicon wafer
- PDMS
- Glass coverslips (N°1–25 mm diameter)
- Holes puncher

4.3.1.2 Equipment

- Plasma cleaner
- Tweezer
- Blade
- Oven

4.3.1.3 Methods

1. Pour PDMS (cross-linker:prepolymer (1:10)) on a silicon wafer. Remove air bubbles by placing the wafer in the desiccator for 1 h.
2. Cure at 65 °C for at least 4 h.
3. Cut PDMS with a blade and remove the PDMS stamp from the wafer (see Fig. 6B (i)).
4. Punch holes in the reservoirs with a 0.75 mm diameter puncher.
5. Use adhesive tape to remove dusts and PDMS residues deposited on the motifs.
6. Activate the coverslip and PDMS stamp with oxygen plasma. Bind PDMS to the coverslip (see Fig. 6A (ii')).
7. Incubate the chip at 65 °C overnight to secure the binding between the PDMS stamp and the glass coverslip.

4.3.2 Cell migration experiment

At this stage, all steps should be performed under a sterile hood with a laminar flow. To place cells in the sample, we follow this protocol.

4.3.2.1 Materials

- NIH3T3 fibroblasts (or other migratory fibroblasts)
- Trypsin–EDTA
- DMEM 4.5 g/L glucose 1% Penicillin–streptomycin
- Leibovitz's L-15 medium
- BCS
- Pipettes
- Petri dishes
- Eppendorf micro-loader (Eppendorf, cat. 5242956003)

4.3.2.2 Equipment

- Laminar flow hood
- Incubator
- Centrifuge
- Epifluorescence microscope with phase contrast and temperature control
- Metallic holder

4.3.2.3 Method

1. Activate the chip with oxygen plasma to make it hydrophilic. Fill in the PDMS chip with L-15 with 10% BCS.
2. Trypsinize a 5 mL Petri dish with adherent cells for 3 min, and add 4 mL of DMEM 10% BCS. Count cells.
3. Centrifuge for 3 min at 500 rpm. Cells will form a pellet. Remove the supernatant and re-suspend cells in L-15 10% BCS at a density of 30 million cells/mL.

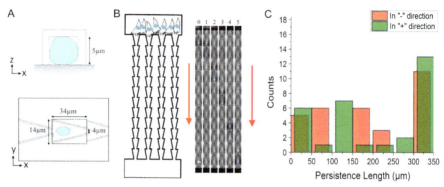

FIG. 8

Dimension of the triangle 3D confined motifs. (A) Schematic of a ratchet unit with its dimensions, *xz* axis (top) and *xy* axis (bottom). (B) Left: Schematics of the experimental setup. Cells are seeded in one side of the channel allowing the entry and migration of the cells in the "+" direction (red arrow). Right: Time-lapse images of a NIH3T3 fibroblast migrating in a ratchet microchannel in the "+" direction (red arrow). Blue dots outline the nucleus over time. Scale bar 10 μm, time in hours. (C) Persistence length distributions of cells migrating in "+" direction and in "−" direction. Experiments in (B) were repeated with cells plated on the other side of the channel. Altogether, the confined configuration forces cells to follow the polarity set at the entry of channels (in preparation).

4. With an Eppendorf micro-loader, inject 20 μL of cells suspension into one inlet (see Fig. 8B). Slowly aspirate L-15 medium from the second inlet to allow accumulation of cells at the entry of channel.
5. Immerse the PDMS chip in L-15 containing 10% BCS.
6. Incubate for 2 h at 37 °C to let cells adhere and spread.
7. Transfer the chip into sterile holder and start the time-lapse acquisition with a 10 × objective and 10 min interval. Note that due to the presence of a thin layer of PDMS in the optical path, phase contrast images might not be optimal. To improve the phase contrast, use an objective with a numerical aperture of a minimum 0.40.

5 CONTROL OF CELL MIGRATION IN 3D WITH CHEMOTAXIS

With this setup, the motion of cells can be challenged by a chemical gradient when the sample is closed (see Fig. 9B). We report now the protocol.

5.1 PROTOCOL

5.1.1 Materials
- Masks/silicon wafer
- PDMS
- Glass coverslips (N°1–25 mm diameter)
- Holes puncher

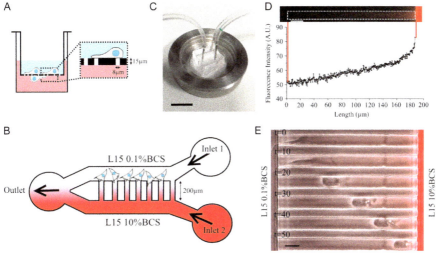

FIG. 9

Boyden and microfluidics experiments to study chemotaxis. (A) Schematic of a Boyden chamber. Typically, a Boyden chamber is composed of two chambers separated by a 15 μm thickness membrane with 8 μm pores (Boyden, 1962). Cells are seeded in the upper chamber in medium without chemo-attractant. The lower chamber is filled with medium containing the chemo-attractant. After incubation for 12 h, migration is stopped and migrating cells which crossed the membrane are counted. (B) Schematic of the microfluidic chip used to follow cell migration in response to a chemical gradient. Cells are seeded in the microfluidics chamber. Gradient is formed in the micro-channels by connecting two syringes, one with the chemo-attractant and fluorescent dye and the other one without. A syringe pump is necessary to form stable gradient up to 24 h. (C) Image of the device. Scale bar 1 cm. (D) Top: Fluorescent image of the microchannel and the chemical gradient visualized with fluorescent dye. Bottom: Corresponding fluorescent intensity profile along the microchannel. Scale bar 20 μm. (E) Time-lapse images of a NIH3T3 fibroblast migrating in a microchannel in a gradient of 10% serum. Scale bar 20 μm, time in minutes.

5.1.2 Equipment
- Plasma cleaner
- Tweezer
- Blade
- Oven

5.1.3 Methods
1. Pour PDMS (cross-linker:prepolymer (1:10)) on a silicon wafer. Remove air bubbles by placing the wafer in the desiccator for 1 h.
2. Cure at 65 °C for at least 4 h.
3. Cut PDMS with a blade and remove the PDMS stamp from the wafer (see Fig. 6A (i)).

4. Punch holes in the inlets–outlet with a 0.75 mm diameter puncher.
5. Use adhesive tape on the motifs to remove any residual dusts and/or PDMS residues.
6. Activate the coverslip and PDMS stamp with oxygen plasma. Bind PDMS to the coverslip (see Fig. 6A (ii)).
7. Incubate the chip at 65 °C overnight to secure the binding between the PDMS stamp and the glass coverslip.

5.2 CELL MIGRATION EXPERIMENT

At this stage, all steps should be performed under a sterile hood with a laminar flow. To place cells in the sample, we follow this protocol.

5.2.1 Materials
- NIH3T3 fibroblasts (or other migratory fibroblasts)
- Trypsin–EDTA
- DMEM 4.5 g/L glucose 1% Penicillin–streptomycin
- Leibovitz's L-15 medium
- BCS
- Petri dishes
- Eppendorf micro-loader
- Tetramethylrhodamine isothiocyanate–Dextran, molecular weight 20 kDa (Sigma, cat. 73766)
- 1 mL syringe (Henke-Sass Wolf Soft-Ject, cat. 5010-A00V0)
- 20Gx1″ needle (Terumo 0.9 × 25 mm Luer)
- 23Gx1″ needle (Terumo 0.6 × 25 mm Luer)
- Tygon tubing (Saint Gobain, cat. AAD04103)

5.2.2 Equipment
- Laminar flow hood
- Incubator
- Centrifuge
- Epifluorescence microscope with phase contrast and temperature control
- Metallic holder
- Syringe pump

5.2.3 Method
1. Activate the chip with oxygen plasma to make it hydrophilic. Fill in the PDMS chip with L-15 with no BCS (0% BCS).
2. Trypsinize a 5 mL Petri dish with adherent cells 3 min, and add 4 mL of DMEM containing 10% BCS. Count cells.
3. Centrifuge for 3 min at 500 rpm. Cells will form a pellet. Remove the supernatant and re-suspend cells in L-15 0% BCS at a density of 30 million

cells/mL. Cells are re-suspended in L-15 0% BCS in order to starve cells prior to stimulation to the serum gradient.

4. With an elongated tip, inject 20 μL of cell suspension into the outlet. Slowly aspirate L-15 medium from inlet 1 to allow accumulation of cells at the channels entry (see Fig. 9B).

5. Immerse the PDMS chip in L-15 0% BCS.

6. Incubate for 2 h at 37 °C to let cells adhere and spread.

7. Transfer the chip into the sample holder.

8. To form the serum gradient, two 1 mL syringes are prepared with L-15 0.1% BCS or L-15 containing 10% BCS and fluorescent dextran, enabling visualization of the chemical gradient.

9. Connect syringes to syringe needles (23Gx1″) and Tygon tubing which have previously been rinsed with L-15.

10. Make sure that there are no bubbles into the tubing during experiment. This can prevent a good flow and detach cells if bubbles enter the microfluidic device (see Troubleshooting: Tr.6; Fig. 10F).

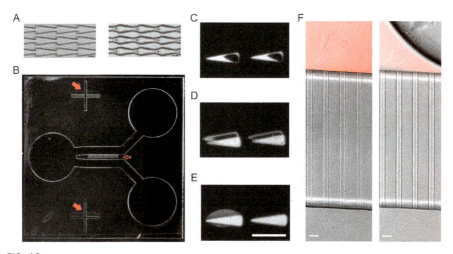

FIG. 10

Issues during microfabrication (A and B), micro-contact printing process (C, D, and E) and microfluidics experiments (F). (A) Images of two wafers. Left: Image of connected ratchet microchannel properly defined. Right: Image of ill-defined connected ratchet microchannel due to poor contact between SU-8 on wafer and mask. Scale bar 20 μm. (B) Image of a wafer where first and second layers are misaligned. Filled red arrows show non-aligned crosses, empty red arrow shows that micro-channels are not connected to the main channel at the bottom side. Scale bar 100 μm. (C) Patterns with over-dried stamp. (D) Double stamping. (E) Wet patterns. (F) Appearance of bubbles during chemical gradient experiment in microchannel. Chemical gradient is removed in the presence of bubble. Scale bar 20 μm.

11. Connect tubing to the two inlets of the chip by using cut syringe needles (20Gx1″). The outer diameter of the needles (0.9 mm) is slightly larger than the diameter of the punched holes (0.75 mm). This allows to make sure that upon insertion the connection is sealed securely (see Fig. 9C).

12. Place syringes on the pump. Flow the two solutions at a rate of 10 μL/h. After few minutes, a stable gradient forms into microchannels which can be visualized with fluorescent dextran as seen in Fig. 9D and E.

Conclusion: With this method, chemical gradients can be generated in closed configurations—but not in open configurations. They can be imposed in competition or in cooperation with other external cues such as confinement and ratchetaxis. The net results can be reported through the analysis of trajectories like in the other examples presented above.

6 TROUBLESHOOTING

Many issues can appear in the preparation of samples, during the microfabrication, micro-patterning, or even during acquisition under the microscope. We follow the order of their appearance below.

6.1 MICROFABRICATION

Tr.1: During microfabrication process, a critical step is to obtain well defined structures, for example, triangles should have sharp edges. Ill-defined structures will appear, if there is not a firm contact maintained between SU-8 and mask (see step in Figs. 3 (iii) and 10A). To prevent this, make sure that the contact is tight enough. This can also happen during the development step when motifs are under or overdeveloped. To avoid this, immerse back the wafer in SU-8 developing solution for 30 s, rinse with isopropanol and check the motifs.

Tr.2: If two layers are required, they need to be perfectly aligned. A misalignment is observed when crosses on the wafer of layer 1 do not overlap with crosses on the mask for the layer 2 (see Fig. 10B filled arrows). This results in disconnected structures from the first and second SU-8 layers (see Fig. 10B empty arrow). To prevent this, be patient and pay attention during alignment step that both crosses are aligned. During the design of the mask, it is possible to insert more asymmetric motifs in both masks (for the layers 1 and 2) along the motifs to make sure that there is optimal alignment, before UV exposure.

6.2 MICRO-CONTACT PRINTING

Tr.3: Drying time critically depends on humidity, the ambient temperature in the room and on patterns characteristics (area, aspect ratio and spacing between motifs). It can vary from 2 min to 6 min depending on the design. Drying the stamps with a nitrogen stream can prevent these problems, but this method is more difficult to control. Drying too much the stamp will impede fibronectin transfer to the glass surface (Fig. 10C). On the other hand, a stamp too wet gives ill-defined patterns (Fig. 10E). This is the most critical step.

Tr.4: Double stamping (Fig. 10D) usually happens when the stamp moves while applying pressure or when it touches again the coverslip while removing it. Using two tweezers to remove the stamp (one on the coverslip, the other one holding the stamp) is a good solution to carefully release the PDMS stamp from the coverslip.

Tr.5: Cell death during acquisition might happen for the following reasons:

— Coverslips are not well rinsed after "piranha" treatment.
— Stability of PLL-g-PEG: it is recommended not to use the diluted solution after 2–3 weeks.
— Cells are too confluent on the plate, prior deposition on patterns: contact inhibition might affect cell motility and prevent motion leading to death on the motifs (especially for large gap distances).

6.3 MICROFLUIDICS

Tr.6: While imaging, air bubbles can affect the stability of chemical gradients (see Fig. 10F), and can lead to detachment of cells from the coverslip. Typically, this type of experiments cannot be analyzed. To prevent this, make sure that there are no bubbles in the tubing while preparing them and before starting experiment.

Tr.7: Even after the sealing of PDMS stamps to the coverslip, sometimes detachment can occur. A proper binding is particularly important for generating the chemical gradient. In this situation, check that plasma binding is properly achieved. Also, check if PDMS and coverslips are clean and without dusts, which if present could prevent proper sealing.

Optimal z-focus: In order to image cells during typical timescales relevant for reliable measures, i.e., days, cells need to remain within the same focal plane. When acquiring for longer time periods at $37\,°C$, microscopes might exhibit some drift due to thermal gradients, in particular along the z-direction. A good solution involves placing the entire experimental set-up in an environment with temperature control ahead of time, prior the experiment to secure a good mechanical stability. Alternatively, feedback loops to control the objective-sample distance can be used to keep focus throughout acquisition. Also, μm size beads can be deposited on the sample, and images can be realigned after experiments using these fiducial markers.

7 CONCLUSION

The methods reported in this article show how cell migration can be controlled specifically for a quantitative study. Their implementations do not require large and expensive equipment and these assays can be used in any laboratories. Designs of 2D and 3D cellular controls have allowed to reveal new modes of migrations

(Bergert et al., 2015; Caballero et al., 2015; Liu et al., 2015). With such setups, more migration modes could appear and the associated signaling networks could be studied. Finally, these microfabrication techniques could also open potential new methods for diagnosis at cellular scales in diseases where migration is impaired, in cancer for example (Mitchell, Jain, & Langer, 2017).

REFERENCES

Bergert, M., Erzberger, A., Desai, R. A., Aspalter, I. M., Oates, A. C., Charras, G., et al. (2015). Force transmission during adhesion-independent migration. *Nature Cell Biology*, *17*(4), 524–529.

Boyden, S. (1962). The chemotactic effect of mixtures of antibody and antigen on polymorphonuclear leucocytes. *Journal of Experimental Medicine*, *115*(3), 453–466.

Caballero, D., Comelles, J., Piel, M., Voituriez, R., & Riveline, D. (2015). Ratchetaxis: Long-range directed cell migration by local cues. *Trends in Cell Biology*, *25*(12), 815–827.

Caballero, D., Voituriez, R., & Riveline, D. (2014). Protrusion fluctuations direct cell motion. *Biophysical Journal*, *107*(1), 34–42.

Comelles, J., Caballero, D., Voituriez, R., Hortigüela, V., Wollrab, V., Godeau, A. L., et al. (2014). Cells as active particles in asymmetric potentials: Motility under external gradients. *Biophysical Journal*, *107*(7), 1513–1522.

Cramer, L. P. (2010). Forming the cell rear first: Breaking cell symmetry to trigger directed cell migration. *Nature Cell Biology*, *12*(7), 628–632.

Hawkins, R. J., Piel, M., Faure-Andre, G., Lennon-Dumenil, A. M., Joanny, J. F., Prost, J., et al. (2009). Pushing off the walls: A mechanism of cell motility in confinement. *Physical Review Letters, 102*, 058103-1–058103-4.

Helvert, S. V., Storm, C., & Friedl, P. (2018). Mechanoreciprocity in cell migration. *Nature Cell Biology*, *20*(1), 8–20.

Jiang, X., Bruzewicz, D. A., Wong, A. P., Piel, M., & Whitesides, G. M. (2005). Directing cell migration with asymmetric micropatterns. *Proceedings of the National Academy of Sciences of the United States of America*, *102*(4), 975–978.

Liu, Y. J., Le Berre, M., Lautenschlaeger, F., Maiuri, P., Callan-Jones, A., Heuzé, M., et al. (2015). Confinement and low adhesion induce fast amoeboid migration of slow mesenchymal cells. *Cell*, *160*, 659–672.

Mitchell, M. J., Jain, R. K., & Langer, R. (2017). Engineering and physical sciences in oncology: Challenges and opportunities. *Nature Reviews. Cancer*, *17*, 659–675.

Petri, J. R. (1887). Eine kleine Modification des Koch'schen Plattenverfahrens. *Centralblatt für Bakteriologie und Parasitenkunde*, *1*, 279–280.

Prentice-Mott, H. V., Chang, C.-H., Mahadevan, L., Mitchison, T. J., Irimia, D., & Shah, J. V. (2013). Biased migration of confined neutrophil-like cells in asymmetric hydraulic environments. *Proceedings of the National Academy of Sciences of the United States of America*, *110*(52), 21006–21011.

Qin, D., Xia, Y., & Whitesides, G. M. (2010). Soft lithography for micro- and nanoscale patterning. *Nature Protocols*, *5*(3), 491–502.

Théry, M., & Piel, M. (2009). Adhesive micropatterns for cells: A microcontact printing protocol. *Cold Spring Harbor Protocols*, *4*(7), 1–12.

Whitesides, G. M. (2006). The origins and the future of microfluidics. *Nature*, *442*, 368–373.

3D cell migration in the presence of chemical gradients using microfluidics

8

Andrew G. Clark*,1, Anthony Simon*, Koceila Aizel†, Jérôme Bibette†, Nicolas Bremond†, Danijela Matic Vignjevic*

*Institut Curie, PSL Research University, CNRS, UMR 144, Paris, France
†Laboratoire Colloïdes et Matériaux Divisés, CNRS UMR 8231, Chemistry Biology & Innovation, ESPCI Paris, PSL Research University, Paris, France
1Corresponding author: e-mail address: andrew.clark@curie.fr

CHAPTER OUTLINE

Methods in Cell Biology, Volume 147, ISSN 0091-679X, https://doi.org/10.1016/bs.mcb.2018.06.007

Abstract

Chemotaxis is an important biological process involved in the development of multicellular organisms, immune response and cancer metastasis. In order to better understand how cells follow chemical cues in their native environments, we recently developed a microfluidics-based chemotaxis device that allows for observation of cells or cell aggregates in 3D networks in response to tunable chemical gradients (Aizel et al., 2017). Here, we describe the methods required for fabrication of this device as well as its use for live imaging experiments and subsequent analysis of imaging data. This device can be adapted to study a number of different cell arrangements and chemical gradients, opening new avenues of research in 3D chemotaxis.

1 INTRODUCTION

Directed cell migration toward chemical cues plays an important role in many biological processes. During immune response, lymphocytes follow chemical signals to reach sites of infection and to locate pathogens (Sarris & Sixt, 2015). In development, cells with different fates migrate in accordance with chemical, among other, cues to reach pre-determined sites to ensure proper body organization (Bussmann & Raz, 2015; Cai & Montell, 2014; Reig, Pulgar, & Concha, 2014). Finally, during metastatic invasion of cancers, tumor cells are thought to be led toward the vasculature by following chemokine gradients (Roussos, Condeelis, & Patsialou, 2011).

A number of recent studies have demonstrated that cells employ different migration mechanisms in 2D and 3D environments. There exists, therefore, a need for new experimental systems that allow testing of the effects of chemical gradients in 3D environments, to better reflect cells' native environments. Currently, most 2D and 3D gradient systems rely on diffusion-based mechanisms using large sink and source reservoirs. However, such gradients are not stable and cannot be used over long periods of time, which is especially important in 3D migration studies. To address this, we recently developed a microfluidics-based chemotaxis device that can be used to expose cells or large aggregates of cells embedded in 3D matrices to chemokine gradients (Aizel et al., 2017). This device can be operated in different modes, where gradients are generated using a diffusion-only or a combination convection–diffusion mechanism to create stable, tunable 3D gradients.

Chemotaxis experiments in 3D cultures using microfluidics can be described in three major steps. First, the gradient device is fabricated. This process involves the design of a silicon master by soft-lithography, fabrication of a PDMS mold, and the subsequent mounting and treatment of the PDMS mold. Second, the cells are embedded in an extracellular matrix and inserted into the chemotaxis chamber and prepared for live imaging. Third and finally, the imaging data is analyzed.

2 GRADIENT DEVICE FABRICATION

2.1 EQUIPMENT

2.1.1. Soft-lithography equipment: photoplotter, hot roll laminator, UV masking system, incubators

2.1.2. Vacuum chamber

2.1.3. Oxygen plasma cleaner

2.2 MATERIALS

2.2.1. 4″ diameter silicon wafers

2.2.2. SUEX thick film dry sheets (500 μm thickness, DJ MicroLaminates)

2.2.3. PDMS (Sylgard 184, Dow Corning)

2.2.4. Aluminum weigh dishes with 4″ inner diameter

2.2.5. Razor blade or scalpel

2.2.6. 1.5 mm diameter biopsy punch

2.2.7. Clear tape (e.g., Scotch)

2.2.8. Aquapel (PGW LLC)/other appropriate silane (see below)

2.2.9. Novec 7500 (3M)/ethanol (see below)

2.2.10. Glass slides (standard size, ca. 75 mm × 25 mm × 1 mm)

2.2.11. Isopropanol

2.2.12. 70% Ethanol

2.2.13. 0.2% Acetic Acid

2.2.14. Rat Tail Collagen-I (Corning)

2.2.15. Poly-D-Lysine

2.3 DESIGN AND FABRICATION OF THE SILICON MASTER

The mold for the chemotaxis chamber can be fabricated using standard soft-lithography techniques. To achieve chamber heights to accommodate large cell aggregates or pieces of tissue, thick film (500 μm) dry sheets should be used instead of spin-coating liquid photoresist. The chamber consists of a 1.4 mm-wide central chamber to house the ECM network and cells and two 1.1 mm-wide lateral flow channels (Aizel et al., 2017) (Fig. 1A–C). The central chamber and flow channels are separated by 200 μm-wide trapezoidal pillars, with a minimal distance of 100 μm between neighboring pillars. For more exact dimensions, a blueprint of the chamber design is included in several file formats (see Supplementary material in the online version at https://doi.org/10.1016/bs.mcb.2018.06.007).

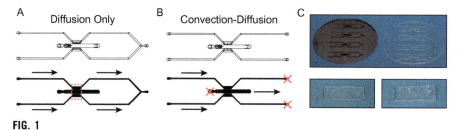

FIG. 1

Chemotaxis chamber design. Technical plans and design of the diffusion-only (A) and convection–diffusion (B) chemotaxis chambers. The blueprints are also included in the Supplementary material in the online version at https://doi.org/10.1016/bs.mcb.2018.06.007. The *red dashed* line in the diffusion-only chamber represents the region that should be exposed to plasma treatment for chemical patterning to make the channels hydrophobic. The red "X"s in the convection–diffusion chamber indicates entry/exit points that are closed with PDMS-filled pipette tips to direct the flow through the central chamber. (C) Top-left: the silicon master for four diffusion-only chambers. Top-right: PDMS mold fabricated using the silicon master. Bottom: cut, punched and mounted diffusion-only (left) and convection–diffusion (right) chambers. PDMS-filled pipette tips are used to block the central chamber entry and flow-channel exits in convection–diffusion mode.

2.4 FABRICATION, TREATMENT AND ASSEMBLY OF THE PDMS CHAMBER

*Convection–diffusion chambers can be fabricated directly using the convection–diffusion pattern; alternatively, the diffusion-only pattern can be used, and these chambers can be optionally converted to convection–diffusion chambers (see below).

2.4.1. Mix PDMS 10:1 with curing agent and degas in a vacuum chamber.
2.4.2. Place the silicon master in an aluminum weigh dish.
2.4.3. Pour the PDMS solution on top of the mold to a height of ~3 mm and bake at 70–80 °C, 4 h to overnight.
2.4.4. Once the PDMS has cured, use a razor blade or scalpel to cut the aluminum weigh boat to extract the master and cured PDMS and carefully peel the PDMS away from the master (Fig. 1C, top).
2.4.5. Cut out blocks of cured PDMS around the chemotaxis chamber using a razor blade.
2.4.6. Punch holes in the entry and exits for the collagen chamber and flow channels using a 1.5 mm diameter biopsy punch.
       *For converting diffusion chambers to convection–diffusion chambers, punch three exit holes corresponding to the normal diffusion exit position and the two corners that correspond to the convection–diffusion exits.

2.4.7. Remove any dust from the channel-side of the chambers using clear tape. Using a piece of transparency film or clear tape, make a mask to cover the chamber such that only the central chamber region (Fig. 1A, *bottom left*, red dashed line) is exposed.

2.4.8. Briefly plasma treat the chamber (channel/mask side up; 2 min) to activate the surface.

*For this plasma treatment, oxygen from the ambient air is sufficient, with a pressure of 800–1100 mTorr in the chamber.

2.4.9. Remove the mask and put the block on a glass slide that has been cleaned with isopropanol (chamber side against glass) and press down firmly to get a temporary seal.

2.4.10. Fill the entire chamber with Aquapel (~20 µL) via the flow channel exit(s).

2.4.11. Incubate 30 s at room temperature.

2.4.12. Aspirate the Aquapel and fill the chamber with Novec 750 to absorb remaining Aquapel.

*As an alternative to Aquapel/Novec 750 treatment, another appropriate silane (e.g., octadecyltrichlorosilane dispersed in ethanol (1 wt%)) can be used, followed by a flush with Ethanol.

2.4.13. Remove PDMS block from glass and clean with isopropanol; dry with compressed air.

2.4.14. Once the PDMS block has dried, test the chamber by sticking it to a clean glass slide, pressing to get a temporary seal, and then loading water in the central collagen chamber (make sure it does not leak into flow channels). If the water does leak into the flow channels, repeat the Aquapel treatment and re-test.

2.4.15. Bond the PDMS blocks (channel-side down) to clean glass slides by oxygen plasma treatment (ideally, assembly should be performed under a laminar flow hood to reduce the risk of dust entry in the chamber). To ensure a tight seal, bake the chambers at 70–80 °C for at least 30 min (Fig. 1C, bottom).

2.4.16. For conversion of diffusion to convection–diffusion chambers, tip the chambers with the exit side down and back-fill the diffusion chamber exit using degassed PDMS/curing agent mixture (as above). Fill using a p20 micropipette until the PDMS just reaches the punched convection–diffusion exits. Bake the chambers at 70–80 °C for at least 30 min.

2.5 COATING OF CENTRAL COLLAGEN CHAMBER

2.5.1. Fill the entire chamber (both the central channel and flow channels) with 70% Ethanol and incubate at room temperature for 5 min.

2.5.2. Aspirate the Ethanol and dry the chambers completely by placing under vacuum for 15 min.

2.5.3. Fill the central collagen chamber with 50 µg/mL collagen-I diluted in 0.2% acetic acid. (The collagen will not polymerize under these acidic conditions.) Incubate for 1 h at 37 °C.

2.5.4. Aspirate the non-polymerized collagen solution and rinse the central chamber with sterile water.

2.5.5. Fill the central chamber with 100 µg/mL Poly-D-Lysine. Incubate for 1 h at 37 °C.

2.5.6. Aspirate the Poly-D-Lysine solution and rinse the central chamber with sterile water.

2.5.7. Repeat the monomeric collagen/Poly-D-Lysine coating procedure two more times, and then rinse the chamber three times with sterile water.

2.5.8. Dry the chambers by aspirating as much water as possible from the central chamber and placing under vacuum for 15 min.

2.5.9. Coated chambers can be stored at 4 °C for ~1 week. However, freshly-coated chambers will better prevent detachment of the collagen gel. For chambers stored longer than one week (but less than one month), repeat the monomeric collagen/Poly-D-Lysine coating procedure once just prior to loading the chambers.

3 CHAMBER PREPARATION FOR LIVE IMAGING EXPERIMENTS
3.1 EQUIPMENT

3.1.1. Syringe pump (e.g., neMESYS) with glass syringes (2.5 mL; SGE)

3.1.2. Vacuum chamber

3.1.3. Fluorescence microscope (epifluorescence is sufficient to observe the gradient)

3.2 MATERIALS

3.2.1. Cells or cell aggregates with appropriate CO_2-independent medium

3.2.2. 50 mL Falcon tubes

3.2.3. Fluorescent dye to visualize gradient

3.2.4. 20 Ga hypodermic needle

3.2.5. 0.8 mm inner/1.58 mm outer diameter PTFE tubing (VWR)

3.2.6. Rat Tail Collagen-I (Corning)

3.2.7. $10 \times$ PBS

3.2.8. 1 N NaOH

3.2.9. Plastic trays (Omni Tray, Thermo Fisher Scientific)

3.2.10. Silicon grease (e.g., High Vacuum Grease, Dow Corning)

3.2.11. Clear tape (e.g., Scotch)

3.3 MEDIUM PREPARATION

3.3.1. Prepare 15 mL of cell media, which will depend on what cell type you are using.

   *It is not possible to use CO_2 to buffer solutions in this setup. To accommodate this limitation, use a compatible CO_2-independent medium, such as L15, or add 20 mM HEPES to your medium for buffering.

   *To avoid contamination in the syringes, it is recommended to use a high dose of antibiotics ($2\times$ Penicillin/Streptomycin) and a broad-spectrum antimicrobial (against bacterial, fungi and mycoplasma) agent (e.g., Primocin, InvivoGen).

3.3.2. Split the medium into two 50 mL Falcon tubes, with 5 mL for chemokine "+" medium and 10 mL chemokine "−" medium.

3.3.3. To the chemokine "+" medium, add desired chemokine as well as a fluorescent marker (e.g., unlabeled Fluorescein or TAMRA, or fluorescently labeled Dextran) to visualize the gradient.

   *To ensure the fluorescent gradient reflects the chemokine gradient, the fluorophore should have a similar size as the chemokine.

3.3.4. With the lids of the Falcon tubes removed, degas the media by placing under a strong vacuum for 2 h

   *To reduce the risk of contamination during degassing, the vacuum chamber should be cleaned with bleach and a microbial disinfectant (e.g., Surfa'Safe, Anios) immediately prior to placing tubes in the vacuum chamber.

3.4 LOADING OF COLLAGEN AND CELL MIXTURE INTO THE CHAMBER

3.4.1. For convection–diffusion experiments, make several PDMS-filled p20/200 pipette tips ahead of time. Cut tips to ∼1–1.5 cm (see Fig. 1C, *bottom right*).

3.4.2. Place chemotaxis chambers under strong vacuum for at least 20 min prior to loading collagen/cells. This will help to remove any air bubbles that form in the chamber during loading.

3.4.3. Prepare 100 μL neutralized collagen solution on ice (see Appendix for collagen solution recipe).

   *For chemotaxis studies in single cells, add suspended cells directly to the collagen mixture (see Appendix).

   *For studies using cell aggregates, aggregates can be formed using a number of methods, including culture in round agarose wells, hanging drop, or alginate capsules. For embedding aggregates, make a second collagen solution. Deposit the aggregates in one of the collagen solutions in a minimal volume of medium. Then transfer the aggregates to the second solution of collagen solution in a minimal volume to preserve the desired collagen concentration.

3.4.4. Inject collagen solution into the central chamber to fill the entire length of the chamber (20 µL is sufficient).

*For large cell aggregates, fill the central chamber with the cell-free collagen, and then immediately inject aggregates from the other collagen solution. Ensure that aggregates are well-spaced (min. 1.5 mm between aggregates) to avoid interaction between aggregates. Then immediately invert the chamber and tilt to position the aggregate(s) correctly in the chamber (this must be done quickly before the collagen polymerizes).

3.4.5. For convection–diffusion experiments, plug the collagen chamber entry with a PDMS-filled p20/200 pipette tip immediately after injecting the collagen.

3.4.6. Place the slide with the chamber in a 20 cm petri dish and fill the dish with just enough water so that the water is in contact with the PDMS block.

*For large cell aggregates, wet two Kimwipes and ball them up. Place the wet balls on opposite sides of a 20 cm Petri dish. Balance the chamber (still inverted) on the balls and place the cover on the Petri dish. The inversion of the chamber during collagen polymerization ensures that the aggregate will not contact the glass slide, which will result in 2D spreading/ migration on the glass surface.

3.4.7. Incubate the chambers in humidified conditions for 20 min, until the collagen polymerizes.

*The topology and mechanical properties of collagen networks are highly dependent on collagen concentration, pH and polymerization temperature (Sapudom et al., 2015; Wolf et al., 2013). For example, lower polymerization temperatures result in longer fibers and larger meshsizes, while higher polymerization temperatures result in shorter fibers and smaller meshsizes.

3.4.8. Once the collagen has polymerized, slowly fill the side channels with the chemokine "−" medium using a p20 micropipette. For diffusion-only chambers, fill the side channels via the single exit hole. For convection– diffusion chambers, simultaneously fill the side channels via the entry holes with two p20 micropipettes.

3.4.9. Place the filled chambers in an incubator with the appropriate temperature for your cells. Remember that the cells should now be in CO_2-independent conditions.

3.5 PREPARATION OF SYRINGES AND SYRINGE PUMP

3.5.1. Clean the syringes by filling with 70% Ethanol and incubating at room temperature for 15 min with the syringes inverted (plungers facing down). Remove the Ethanol and allow the syringes to air dry for at least 1 h.

3.5.2. Measure out the required length of the tubing from the syringe pump to the chamber (this will depend on your microscope setup). File down the sharp tip of a 20 Ga hypodermic needle and insert it into one end of the PTFE tubing (Fig. 2A). Cut the opposite end of the tube at a 45° angle to achieve a sharp tip at the end of the tubing to aid insertion into the PDMS block. The needle/tubing setup can be reused for future experiments.

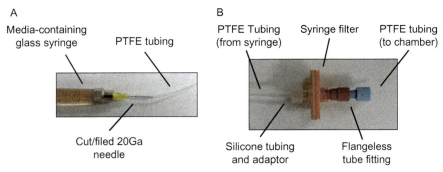

A
Media-containing glass syringe   PTFE tubing

Cut/filed 20Ga needle

B
PTFE Tubing (from syringe)   Syringe filter   PTFE tubing (to chamber)

Silicone tubing and adaptor   Flangeless tube fitting

FIG. 2

Syringe tip and bubble trap arrangements. (A) A glass syringe is connected to PTFE tubing using a 20 Ga hypodermic needle that has been cut/filed to remove the sharp tip. (B) Bubbles can be trapped immediately before flow into the chamber using a standard syringe filter. Here, the PTFE tubing from the syringe is attached to the filter using a short length of Tygon silicone tubing and a plastic tubing adaptor. Downstream of the filter, a short length of PTFE tubing (to be inserted into the chamber) is attached using a flangeless tube fitting.

3.5.3. Clean the assembled tubes/needles by filling with 70% Ethanol and incubating at room temperature for 2 h. Flush out the Ethanol with compressed air and allow to air dry at least 10 min.

3.5.4. Very carefully fill the syringes from the back using a p1000 micropipette with the chemokine "−" and "+" media. Cover the nozzle with your (gloved) finger to prevent the syringe from leaking, and carefully insert the plunger. Invert the syringes and remove any small bubbles by tapping with your finger and pushing the bubbles out using the plunger.

3.5.5. Fill a plastic syringe with the remaining volume of chemokine "−" medium.

3.5.6. Fill the tubes with 70% Ethanol from the needle attachment using a plastic syringe. To prevent the Ethanol from exiting the tube entry, hold the exit above the needle. The Ethanol will help the medium to wet to the tube, preventing formation of air bubbles during filling.

3.5.7. Replace the Ethanol in the tubes using the plastic syringe filled with chemokine "−" medium, being sure to leave an excess of medium on the needle opening.

3.5.8. Carefully attach the glass syringe to the tube, applying a little pressure to the plunger to prevent air bubbles from forming during attachment. From this point, the syringes should remain upright (plunger-side up).

*For long term experiments, it is recommended to install a bubble trap just upstream of the chamber entry point to prevent air bubbles from entering the chamber and disrupting the collagen and gradient (Fig. 2B).

*As an alternative to syringe pumps, a microfluidic pressure controller can be used.

3.6 MOUNTING OF CHAMBER AND FLOW TUBES

3.6.1. Mount the chamber on a clear plastic tray, sealing the bottom edges with silicon grease and clear tape. Add enough water to the tray so that the water just contacts the PDMS block. Mount the tray on the microscope (Fig. 3).

3.6.2. Cover the tray with a lid that has been modified to allow for insertion of the syringes (Fig. 3B, inset). A hole can be cut in the plastic lid using a hot scalpel or a drill.

3.6.3. Mount the syringe pump near the microscope on its side so that the syringe plungers will be facing up (Fig. 3). This ensures that any bubbles nucleating in the syringes will not enter the tubes.

    *Ideally, the syringe pump should be housed inside of a large heating chamber with the microscope to ensure that the medium stays at 37° (for experiments using mammalian cells). In addition, reducing temperature fluctuations in the medium will reduce the risk of air bubble nucleation.

3.6.4. Attach the syringes to the syringe pump and ensure you have flow from both tubes.

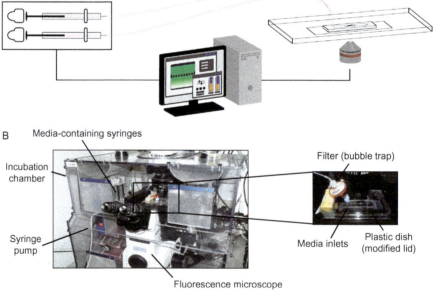

FIG. 3

Microfluidics setup for live cell imaging. (A) A syringe pump is used to deliver a precise and stable flow of media from glass syringes. The syringes are connected to the flow entry points of the chamber using flexible tubing. Both the syringe pump and the microscope can be controlled by a single computer. (B) The microfluidic setup mounted on a fluorescence microscope for cell imaging. The syringe pump is housed inside of the microscope's heating chamber. The flow of media from the syringes can be optionally passed through a bubble trap before connection to the chamber, which is mounted on the microscope for live imaging.

3.6.5. Set the flow rates to 50 μL/h and insert the tubes in the side channel entries. Ensure that you have flow in both channels by viewing the channels in the microscope. You should see flow of small particulate in the medium in both flow channels as well as the fluorescent marker in one side channel (the chemokine "+" side).

3.6.6. For convection–diffusion gradients, once flows have been established, stop the pumps and wait 3 min. Then very gently insert PDMS-filled pipette tips in the flow channel exits.

3.6.7. Set the flows to 20 μL/h.

3.6.8. Place a thin piece of Kimwipe just upstream of the exit hole(s). This will wick the medium away, preventing accumulation on top of the chamber (which can interfere with imaging). Alternatively, a shallow slit just upstream of the exit hole(s) can be made in the PDMS using a razor blade, and a thin piece of coverslip can be inserted into the slit. (This should be done prior to loading the ECM/cells.) This will prevent the exiting medium from accumulating over the imaging area and interfering with imaging.

3.7 LIVE IMAGING

Live imaging parameters will depend on the speed of the cells and whether cells are fluorescently labeled. Even without fluorescent labeling, pre-processing of bright-field images (below) can ensure high-quality cell tracking. Time steps may range from 2 min to 15 min, with z-intervals of 5–50 μm. Even with low magnification objectives (e.g., $10\times$), the microscope field will be considerably smaller than the chemotaxis chamber. In this case, it is necessary to take a grid of several stage positions, which can be stitched together during preprocessing. On some microscope setups, such grid stitching may be automatic, while on other setups, the stage positions must be set manually. With the described setup, chemotaxis experiments can be conducted over several days.

---

4 DATA ANALYSIS
4.1 PREPROCESSING

Preprocessing can most easily be performed using ImageJ/FIJI (Schindelin et al., 2012), which has a wealth of available plugins and features for image processing and is easily amenable to batch analysis using scripts written in the ImageJ Macro language, Java or other compatible programming languages. Icy (de Chaumont et al., 2012) is an alternative open-source image analysis platform that is scriptable and offers a wealth of community-supported plugins. Below is a sample preprocessing workflow for 2D tracking of cells from a 3D chemotaxis experiment.

4.1.1. Compile multi-color image stacks for each stage position.

4.1.2. Stitch positions using the Grid/Stitching plugin (Preibisch, Saalfeld, & Tomancak, 2009).

4.1.3. Rotate and crop the image stack appropriately.

4.1.4. Compress all frames onto a single plane using the Extended Depth of Field plugin (Forster, Van De Ville, Berent, Sage, & Unser, 2004). This step should be avoided if you plan to track cells in 3D.

4.1.5. Highlight the cell positions using a variance filter followed by a Gaussian filter.

4.1.6. Register the time series using the TurboReg plugin (Thevenaz, Ruttimann, & Unser, 1998).

4.2 CELL TRACKING AND ANALYSIS

Cell tracking can be performed using a number of different software packages, including ImageJ/FIJI, Imaris, Ilastik or Icy. Depending on the specifics of your images, you may choose a manual, automated or semi-automated approach. For analysis of cell aggregates, single escaping cells can be tracked individually, and the morphodynamics of the cluster itself can be analyzed using the above software packages.

In chemotaxis studies, the most interesting characteristic of a cell's trajectory is its direction. Commonly, this is quantified by the chemotaxis index (CI). The CI is best defined as displacement along the gradient divided by the total trajectory length (Othmer, Dunbar, & Alt, 1988). $CI = 1$ therefore represents migration straight up the gradient, while $CI = -1$ represents migration straight down the gradient. Alternatively, CI has been defined as $CI(i) = \cos(\phi - \theta_i)$, where ϕ represents the angle directly up the gradient and θ_i is the angle between the start and end points of each track (Kay, Langridge, Traynor, & Hoeller, 2008). Although this second definition accounts for the direction change from start to end, it does not account for the total track length and can be highly dependent on which timepoints are used for tracking. The first definition of CI is therefore preferred.

In addition to CI, other information related to cell speeds or morphology may also be desired. These data can often be extracted directly from the tracking software or from the cell trajectories.

5 COMMON EXPERIMENTAL PROBLEMS AND TROUBLESHOOTING

- Detachment of the collagen gel (during media perfusion, insertion of flow tubes or over the course of imaging)
  - Be sure monomeric collagen/Poly-D-Lysine coating is fresh.
  - While permeating the side channels, pipette very slowly. If using two pipettes (for convection–diffusion chambers), make sure the fluid fronts move at the same speed.

- Ensure the tip of the flow tube is cut at an angle to allow easy insertion into the PDMS block, and insert the tube very slowly.
- Insert the second flow tube immediately after the first is inserted.
- If the collagen detaches while inserting the PDMS-filled pipette tips, be sure that the flow is indeed stopped (by watching for particulate flow in the microscope) for a full 3 min before inserting. This will prevent the back-flow that causes detachment.
- Before starting imaging, make sure that the gel is still tightly attached to all pillars.
- Appearance of bubbles in flow channels
  - Ensure media is well degassed, and fill and close the syringes immediately after degassing media.
  - Be careful not to introduce bubbles at the syringe/needle junction during assembly.
  - Inspect the needle/tube for bubbles before inserting the tube into the PDMS block. Depress the plunger manually to flush out any bubbles in the tube. If bubbles remain, disassemble the tube and flush again with Ethanol before re-filling.
  - Insert bubble traps just upstream of the chamber entry points.
- Uneven/aberrant gradient
  - Ensure that you have sufficient and equal flow from both tubes before inserting, by inspecting formation of media bubbles at the ends of tubes.
  - Ensure that you have flow through the channels by looking in the microscope for particulate flow through the flow channels and be sure that the collagen has not detached from the pillars or walls of the chamber.
  - Be sure that exits are free from blockage.
  - Examine the glass syringes for leakage out the back (between the plunger and glass walls). If this occurs, try flattening and smoothing the plunger with a hot knife to remove any imperfections. If leakage persists, the plunger/syringe must be replaced.

6 CONCLUSION

Microfluidic-based chemotaxis systems offer flexibility and stability for studies of gradient response during 3D cell migration. The methods presented in this protocol can be extended to a number of different cell systems and contexts. In this system, gradients can be modulated on fast time scales in order to test, for example, how cells respond to fluctuating or changing gradients. In addition, the collagen matrix can also be modulated (either homogeneously, or under gradient conditions) to further explore the combined effects of mechanical and chemical signaling. Such setups provide an ideal experimental platform for testing a number of outstanding questions concerning cell migration and behavior.

APPENDIX. EXAMPLE COLLAGEN RECIPE

Component	Stock Concentration	Final Concentration	Volume (No Cells)	Volume (With Cells)
PBS	10×	1×	10 μL	10 μL
NaOH	1 N	(Variable)[a]	1 μL	1 μL
Cell medium	–		39 μL	19 μL
Cells	2,000,000 mL^{-1}	400,000 mL^{-1}	–	20 μL
Collagen (in 0.2% acetic acid)	4 mg/mL	2 mg/mL	50 μL	50 μL
Total			100 μL	100 μL

[a]*The NaOH volume depends on the amount of collagen added in order to neutralize the acetic acid in the collagen mixture. The volume of 1 N NaOH should be 1:50 of the collagen volume.*

REFERENCES

Aizel, K., Clark, A. G., Simon, A., Geraldo, S., Funfak, A., Vargas, P., et al. (2017). A tuneable microfluidic system for long duration chemotaxis experiments in a 3D collagen matrix. *Lab on a Chip, 17*, 3851–3861.

Bussmann, J., & Raz, E. (2015). Chemokine-guided cell migration and motility in zebrafish development. *The EMBO Journal, 34*, 1309–1318.

Cai, D., & Montell, D. J. (2014). Diverse and dynamic sources and sinks in gradient formation and directed migration. *Current Opinion in Cell Biology, 30*, 91–98.

de Chaumont, F., Dallongeville, S., Chenouard, N., Hervé, N., Pop, S., Provoost, T., et al. (2012). Icy: An open bioimage informatics platform for extended reproducible research. *Nature Methods, 9*, 690.

Forster, B., Van De Ville, D., Berent, J., Sage, D., & Unser, M. (2004). Complex wavelets for extended depth-of-field: A new method for the fusion of multichannel microscopy images. *Microscopy Research and Technique, 65*, 33–42.

Kay, R. R., Langridge, P., Traynor, D., & Hoeller, O. (2008). Changing directions in the study of chemotaxis. *Nature Reviews Molecular Cell Biology, 9*, 455.

Othmer, H. G., Dunbar, S. R., & Alt, W. (1988). Models of dispersal in biological systems. *Journal of Mathematical Biology, 26*, 263–298.

Preibisch, S., Saalfeld, S., & Tomancak, P. (2009). Globally optimal stitching of tiled 3D microscopic image acquisitions. *Bioinformatics, 25*, 1463–1465.

Reig, G., Pulgar, E., & Concha, M. L. (2014). Cell migration: From tissue culture to embryos. *Development, 141*, 1999–2013.

Roussos, E. T., Condeelis, J. S., & Patsialou, A. (2011). Chemotaxis in cancer. *Nature Reviews Cancer, 11*, 573–587.

Sapudom, J., Rubner, S., Martin, S., Kurth, T., Riedel, S., Mierke, C. T., et al. (2015). The phenotype of cancer cell invasion controlled by fibril diameter and pore size of 3D collagen networks. *Biomaterials, 52*, 367–375.

Sarris, M., & Sixt, M. (2015). Navigating in tissue mazes: Chemoattractant interpretation in complex environments. *Current Opinion in Cell Biology, 36*, 93–102.

Schindelin, J., Arganda-Carreras, I., Frise, E., Kaynig, V., Longair, M., Pietzsch, T., et al. (2012). Fiji: An open-source platform for biological-image analysis. *Nature Methods, 9*, 676.

Thevenaz, P., Ruttimann, U. E., & Unser, M. (1998). A pyramid approach to subpixel registration based on intensity. *IEEE Transactions on Image Processing, 7*, 27–41.

Wolf, K., te Lindert, M., Krause, M., Alexander, S., te Riet, J., Willis, A. L., et al. (2013). Physical limits of cell migration: Control by ECM space and nuclear deformation and tuning by proteolysis and traction force. *Journal of Cell Biology, 201*, 1069–1084.

Microfluidics for cell mechanics

Microfluidics for cell sorting and single cell analysis from whole blood

Ramanathan Vaidyanathan*, Trifanny Yeo*, Chwee Teck Lim*,†,‡,1

*Department of Biomedical Engineering, National University of Singapore, Singapore, Singapore

†Mechanobiology Institute, National University of Singapore, Singapore, Singapore

‡Biomedical Institute for Global Health Research and Technology, National University of Singapore, Singapore, Singapore

1Corresponding author: e-mail address: ctlim@nus.edu.sg

CHAPTER OUTLINE

Methods in Cell Biology, Volume 147, ISSN 0091-679X, https://doi.org/10.1016/bs.mcb.2018.06.011

Abstract

The complexity and dynamic evolution of cancer often result in tumor subpopulations containing distinctly heterogeneous cells. During metastasis, these also give rise to heterogeneous circulating tumor cells (CTCs) which are considered to be a hematogenous dissemination from the primary tumor. CTCs represent a viable less-invasive sampling opportunity, also known as liquid biopsy. However, current technological platforms that analyze entire CTC population are not effective due to cell-to-cell variability within the same population and this can manifest differences in genomic expression, cell cycle stages and eventually cellular responses to drug treatments. Here, we present a novel microfluidic approach that involves combination of two microfluidic chips operating under inertial fluid forces and hydrodynamic focusing to rapidly isolate and selectively retrieve bulk as well as single CTCs from whole blood for downstream single cell analysis. It is envisioned that this combinational approach to retrieve single CTCs can cater to several applications including more accurate disease diagnosis as well as formulation of personalized therapeutic strategies.

1 INTRODUCTION

Cancer is regarded an evolutionary process rather than a condition, where tumor microenvironment and dynamic cellular changes over time are constantly resulting in different subpopulations of cells among systemically spread cancer cells (Greaves & Maley, 2012; Klein, 2013). With intra-tumor heterogeneity being widely recognized (Gerlinger et al., 2012) as one of the key traits of metastatic spread, there is great promise for targeted therapies that exploit specific molecular characteristics of the cancer cells (Luo, Solimini, & Elledge, 2009). However, heritable genetic and epigenetic changes as well as phenotypic plasticity of cancer cells often result in acquired drug resistance. Moreover, several lines of disease treatment inevitably result in cancer cells increasingly disparate from the primary tumor that was surgically removed long time ago. Thus, primary tumors are considered as less potent source to retrieve systemic information on cancer spread and associated clonal evolution. Recent evidences suggest the possibility of even localized tumors without clinically apparent metastasis giving rise to breakaway cells (Alix-Panabieres & Pantel, 2013; Cristofanilli et al., 2004) widely known as circulating tumor cells (CTCs).

Predominantly explored as a "liquid biopsy" source, CTCs avoid the need for repeated intra-lesional bioptic sampling that is rarely tolerable for the patients and often not feasible (Cristofanilli et al., 2004, 2005). Subsequent downstream molecular CTC analysis could help uncover traits of cancer cells that can be selected under therapy early on or enable physicians to rapidly adapt treatment strategies (Hayes et al., 2006; Pantel, Brakenhoff, & Brandt, 2008). Over the years, a growing number of technically diverse platforms were developed for CTCs analysis and have overcome impediments associated with isolation and enumeration. However, some of these platforms being developed based on affinity based isolation (Adams et al., 2008;

Gleghorn et al., 2010; Nagrath et al., 2007; Saliba et al., 2010; Stott et al., 2010) continue to face persistent challenges due to intra-tumor heterogeneity between retrieved cancer cells (Spizzo et al., 2004). This leads to the risk of missing out on a number of potential CTC candidates and certain subpopulations due to epithelial-to-mesenchymal transition (EMT) and the resulting down regulation (Pecot et al., 2011) of commonly used epithelial markers (e.g., EpCAM). These limitations have paved way for the development of label-free microfluidic CTC isolation techniques based on bio-physical characteristics such as size (Hyun, Kwon, Han, Kim, & Jung, 2013; Warkiani et al., 2014), density (Gertler et al., 2003), deformability (Hur, Henderson-MacLennan, McCabe, & Di Carlo, 2011) and dielectric properties (Huang et al., 2013). However, integration of most of these platforms in clinical settings is limited by several factors including low throughput, clogging, low recovery, and possible loss of cell viability. Further, with evidences on intra-tumor and inter-cellular heterogeneity suggesting a higher level of biological complexity, it is imperative that CTCs be analyzed at a single cell level to obtain a detailed profile of disease spread (Bendall & Nolan, 2012; Blainey & Quake, 2013). Heterogeneity among CTCs is significant, and cell-to-cell variations occur even within a single blood draw. However, two major obstacles have limited single CTC analysis, first the isolation of individual tumor cells to purity without contaminating white blood cells (WBCs) and second, the ability to comprehensively analyze single cell genomes or phenotypes for diagnostic purposes (Polzer et al., 2014; Powell et al., 2012).

To address these challenges and enable specific selection of desired bulk CTCs and/or single CTC populations, we describe a method that involves an integrated microfluidic platform for complete separation of CTCs followed by isolation of single cells. This versatile technique is based on a combination of two individual microfluidic chips that utilize: (*i*) inertial microfluidic-based separation of CTCs from blood with higher separation resolution using a spiral microchannel device (Hou et al., 2013; Warkiani et al., 2016), followed by (*ii*) hydrodynamic focusing to efficiently trap single cells and providing flexibility to selectively separate any cells of interest (Yeo et al., 2016). We previously developed the spiral microfluidic chip as an ultra-high-throughput size-based separation method for CTC separation, isolation and retrieval from blood. This method takes advantage of the distinct focusing positions of larger CTCs away from the smaller blood cells as a consequence of combined inertial lift and Dean drag forces in a spiral microfluidic device, enabling rapid and continual isolation of viable CTCs (Fig. 1A). The simplicity and robustness in device operation, along with its high throughput, ensures the feasibility of using our CTC separation method in a clinical setting. With the spiral microfluidic chip able to achieve up to 4-log WBC depletion within 1 h (for 7.5 mL of blood), isolated bulk CTCs can be processed through our single cell capture chip to selectively isolate and retrieve individual CTCs for downstream analysis (Fig. 1B). The single cell capture chip utilizes hydrodynamic focusing to usher cells in the flow and passively trap them in individual control chambers alongside the main channel. This capability enabled efficient trapping of single cells and their subsequent retrieval for single cell analysis to analyze the inherent heterogeneity between cancer cells.

Cell Sorting and CTC Isolation

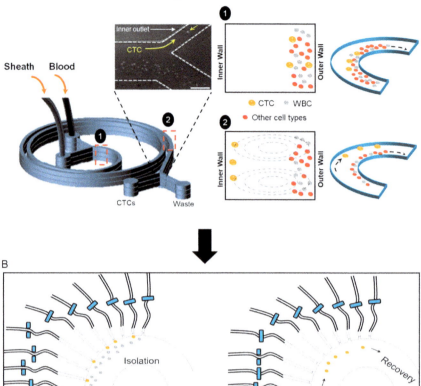

Single CTC Isolation and Recovery

FIG. 1

CTC isolation using spiral and single cell capture microfluidic chips. (A) Schematic illustration of the separation of CTCs using Dean Flow Fractionation (DFF). Blood sample and sheath fluid are pumped into the outer and inner inlets of the spiral microfluidic device, respectively. Under the influence of Dean drag forces, the smaller hematologic cells (RBCs and leukocytes) migrate along the Dean vortices toward the inner wall, then back to outer wall again, while the larger CTCs experience larger inertial lift forces and focus along the microchannel inner wall, thus achieving separation. (B) Single CTC isolation using hydrodynamic focusing of cell flow stream by sheath flow. Cells preferentially enter a chamber due to a net force toward the chamber.

Images were adapted and reproduced with permission from Hou, H. W., Warkiani, M. E., Khoo, B. L., Li, Z. R., Soo, R. A., Tan, D. S.-W., et al. (2013). Isolation and retrieval of circulating tumor cells using centrifugal forces. Scientific Reports, 3, *1259. https://doi.org/10.1038/srep01259; Yeo, T., Tan, S. J., Lim, C. L., Lau, D. P. X., Chua, Y. W., Krisna, S. S., et al. (2016). Microfluidic enrichment for the single cell analysis of circulating tumor cells.* Scientific Reports, 6, *22076. https://doi.org/10.1038/srep22076. https:// www.nature.com/articles/srep22076#supplementary-information. Copyright Nature Publishing Group.*

We envision that the combination of these two microchips in parallel can leverage several applications in cancer diagnosis and monitoring by facilitating a better understanding on CTC heterogeneity. This could be of particular importance during disease monitoring by providing vital information on drug resistance over the course of treatment as well as guide the clinician for better personalized treatment decisions.

2 CELL SORTING AND CTC ISOLATION USING SPIRAL MICROFLUIDIC CHIP

The spiral microfluidic chip is a simple label-free approach that exploits the physical characteristics of different cell types in blood (CTCs ~10–20 μm; RBC ~8 μm discoid; leukocytes ~7–12 μm) to achieve efficient CTC isolation. When these cells migrate across streamlines in curvilinear channels, they tend to focus at distinct positions due to significant inertial forces (henceforth known as inertial microfluidics) which has been exploited for high throughput, size-based cell separation. This cell migration and focusing occurs due to the superposition of two inertial lift forces (F_L) acting on the particles in opposite directions: the shear-induced lift force and the wall-induced lift force (Di Carlo, Irimia, Tompkins, & Toner, 2007). Further, in curvilinear channels cells experience additional lateral Dean drag force (FD) due to the presence of transverse Dean flows arising from the centrifugal acceleration of fluid flow in curved channels. We exploit this interplay between inertial lift and Dean forces for efficient and high resolution size-based separation. The microchannel design consists of a 2-inlet, 2-outlet spiral microchannel (500 μm (w) × 160 μm (h)) with cell focusing being highly reliant on the shortest channel dimension (microchannel height, h) (Bhagat, Kuntaegowdanahalli, & Papautsky, 2008). These channel dimensions aid the larger CTCs in undergoing inertial focusing, while the smaller hematologic cells (RBCs and leukocytes) are only affected by the Dean drag force. When blood sample is driven through the chip, all the cells (CTCs and blood cells) initiate migration along the Dean vortex and move toward the inner channel (Fig. 1A). CTCs focus well near the inner wall due to strong inertial lift forces that minimize any migration under Dean drag force, while blood cells continue flowing along the Dean vortex and travel back toward the outer wall. This enables efficient separation of CTCs at the inner outlet and as shown in Fig. 1A, blood cells (RBCs and leukocytes) are removed from the outer outlet as waste. CTCs collected via the inner outlet will then be utilized to isolate individual CTCs using the single cell capture chip described in Section 3.

Design considerations for spiral microfluidic chip: The spiral chip design involves the use of a simple two-inlet, two-outlet spiral that can enable focusing of different cell types in order to isolate CTCs from lysed blood. The primary design factor for such devices is to devise a spiral microchannel with appropriate channel depth to facilitate inertial focusing of only large target cells near the inner wall while streamlining other smaller cells toward the outer wall. Previous optimization experiments based on investigations on bead behavior in spiral microchannels indicated tight focusing along the inner wall (Kuntaegowdanahalli, Bhagat, Kumar, & Papautsky, 2009). Thus to separate CTCs ($\geq 15\,\mu m$) from other blood cells, channel depth ranging from ~ 150–$180\,\mu m$ was identified to be ideal (Fig. 2). With inertial focusing of large cells occurring at Reynolds number of ~ 20–100, it was important to determine the ideal channel length to achieve suitable output ratios and efficient separation. In this regard, the inner split microchannel for sample inlet was determined to be $75\,\mu m$ while the sheath inlet was around $425\,\mu m$ so that the smaller width of sample inlet forces all the cells entering the spiral channel to begin their lateral migration at a similar position (Fig. 2). Similarly, the split design for channel outlet was determined upon understanding the lateral location of cells across the channel width under different flow rates. Thus, the flow rate required for larger CTCs to focus near the inner wall and the combined sets of flow rate and channel length required for smaller blood cells to travel one complete dean cycle was determined. Under the flow rate and channel length window satisfying both requirements, we measured the distribution of different cell types across the channel width under different input cell concentrations. We then identified the optimal position of outlet split to be $150\,\mu m$ wide for CTC collection outlet near the inner wall and $350\,\mu m$ wide for waste outlet near the outer wall (Hou et al., 2013). Finally, to substantially increase the throughput of our system while simplifying operation and automation, we included an RBC lysis step for improved yield and target cell purity. RBC lysis step substantially reduces the amount of untargeted blood cells in the sample and thus mitigates the undesired cell dispersion due to cell–cell interaction (7.5 mL in 37.5 min for single chip).

2.1 MATERIALS

- Photoresist AZ9260 (for photolithography, Microchemicals)
- Photoresist developer (AZ400K; for photolithography; Microchemicals)
- Isopropyl alcohol (for photolithography)
- PDMS, Sylgard 184 silicone elastomer kit (Dow Corning, Ellsworth Adhesives, cat. no. 184 Sil Elast kit, 0.5 kg)
- Trichloro(1*H*,2*H*,2*H*,2*H*-perfluorooctyl)silane (for soft lithography, silanizing the surface of newly made silicon master; Sigma-Aldrich, cat. no. 448931-10G)
- Sterile PBS, $1\times$ (Corning, cat. no. 21-040-CV)
- BSA, 10% (wt/vol) (Miltenyi Biotec, cat. no. 130-091-376)
- High-glucose DMEM (Life Technologies, cat. no. 11965-092)
- FBS (Life Technologies, cat. no. 10313-039)
- Penicillin-streptomycin (Life Technologies, cat. no. 15140-122)
- Mili-Q purified water
- Trypsin, 0.25% (wt/vol) with EDTA (Lonza, cat. no. CC-5012)
- Cells of interest, e.g., MCF-10, MDA-MB-231, A549 (ATCC, cat. no. HTB-22, CRM-HTB-26, CCL-185)
- RBC lysis buffer, $3\times$ (G-Bioscience, cat. no. 786650)
- EDTA Vacutainer tube (BD Vacutainer, cat. no. 366643)

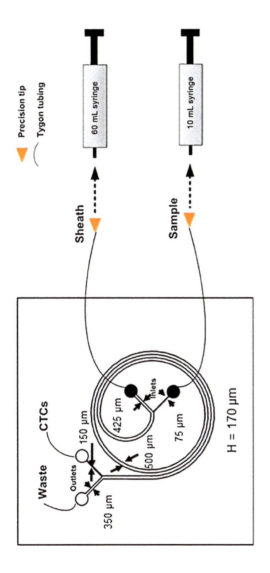

FIG. 2

Design of Spiral microfluidic chip. Schematic representation of spiral microfluidic chip with key design parameters.

- Fixative for counting of CTCs consisting of paraformaldehyde (Sigma-Aldrich, cat. no. P6148)
- Appropriate conjugated antibodies. Fluorescein isothiocyanate (FITC)-conjugated pan-cytokeratin (pan-CK) antibody (1:100; Miltenyi Biotec Asia Pacific) and allophycocyanin (APC)-conjugated CD45 antibody (1:100; Miltenyi Biotec Asia Pacific) for identifying putative CTCs.
- 0.1% (vol/vol) Triton X-100 (Thermo Scientific, cat. no. 85111)
- Hoechst dye (Life Technologies, cat. no. 33342), if nuclei staining is required

2.2 PROCEDURE

2.2.1 Mold preparation for spiral microfluidic chip

Note: All the mold fabrication steps (Fig. 3) must be performed in a designated clean room.

1. Dispense ~6 mL of AZ9260 photoresist onto a cleaned and dehydrated 4-in. wafer.
2. Spin-coat the dispensed AZ9260 film uniformly on the wafer using a spin coater by first ramping to 500 rpm at 100 rpm/s acceleration for 10 s and then ramping to 3000 rpm at 500 rpm/s acceleration for 30 s.
3. Prebake the coated wafer for 5 min at 110 °C.
4. Use a feature-printed transparent film mask to produce the master by exposing the wafer to an appropriate dose of UV light.
5. Develop the exposed wafer in a photoresist developer (AZ400K) for 5 min (1:4 dilution).

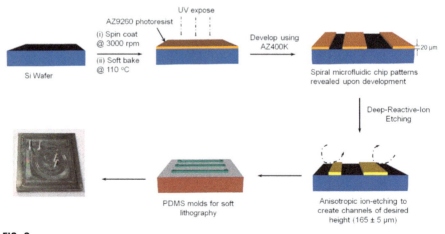

FIG. 3

Fabrication of Mold. Schematic illustration of various steps involved in mold fabrication of spiral microfluidic chip.

6. Rinse the substrate thoroughly with deionized water to stop the development process, and dry the wafer with a gentle stream of pressurized nitrogen gas.

7. Use a DRIE machine to etch the silicon wafer for $\sim 165 \pm 5\,\mu m$. The correct etching depth is crucial for optimum performance of the spiral microfluidic chip. If the channel depth is $<160\,\mu m$ or $>170\,\mu m$, the particle focusing will be compromised, thus causing substantial contamination of the enriched sample.

8. Use a vacuum desiccator to silanize the silicon wafer with $150\,\mu L$ of trichloro ($1H,2H,2H,2H$-perfluorooctyl)silane for 1.5 h. *Note: For reuse of the wafer for subsequent fabrication, it must be stored under a cover to prevent dust exposure for extended periods.*

2.2.2 Production of spiral microfluidic chips (soft lithography)

9. Mix PDMS base with PDMS curing agent homogeneously at a weight ratio of 10:1. With the silanized wafer master secured (using a double-sided adhesive) in the center of a 15-cm-diameter Petri dish, pour 77 g of the PDMS mixture into the dish.

10. Degas the PDMS in a vacuum desiccator for $\sim 1\,h$. When the PDMS is bubble-free, bake the dish with master at 70–80 °C inside the oven for at least 2 h to cure the PDMS.

11. Cut and peel the cured PDMS from the master. The production of spiral microfluidic chip is outlined in Fig. 4. *Note: The cured PDMS can be kept in a clean environment for extended periods and has to be bonded to a glass slide immediately.*

12. Punch holes (1.5 mm dia.) to create inlets and outlets of the device at appropriate points for each PDMS piece.

13. Clean all the PDMS surfaces and glass slide with Scotch tape by gentle tapping. Ensure that the entire area is in contact with the tape, and check for any dirt visually.

14. Plasma-clean immediately the PDMS pieces with features and the glass slide with surfaces to be bonded facing up for 2.5 min at 300 W RF power and $760-10^{-3}$ Torr pressure.

15. Bond the featured PDMS piece to the glass slide by bringing the bonding faces into contact, and press the device softly for 30 s with a pair of tweezers to complete the bonding.
    Note: To avoid channel collapse, do not apply pressure on top of the microchannel feature directly.

16. Place the bonded PDMS device on an 85 °C hot plate or oven for 30 min to further strengthen the bonding.

17. Allow cooling for 5 min and store the chips in a clean environment before further use. Inspect the chips by cutting them across different sections of the chip and measuring the channel width and height using a microscope. This is particularly important as these are key parameters for cell focusing.

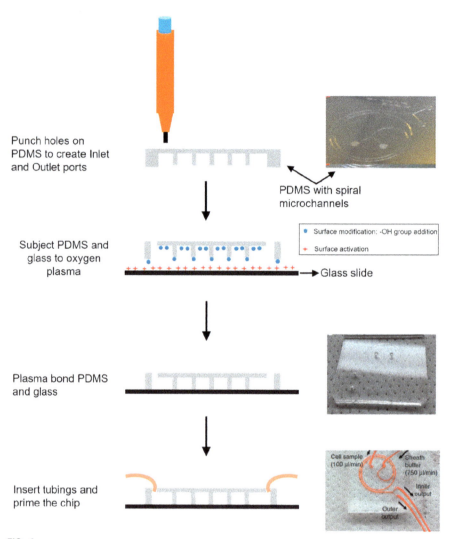

Punch holes on PDMS to create Inlet and Outlet ports

PDMS with spiral microchannels

Subject PDMS and glass to oxygen plasma

Glass slide

Plasma bond PDMS and glass

Insert tubings and prime the chip

FIG. 4

Production of spiral microfluidic chip. Schematic illustration of steps involved in spiral microfluidic chip production from PDMS fabrication through to plasma bonding for final chip production.

2.2.3 Priming of spiral microfluidic chips

18. Connect tubings to the inlets and outlets of the microfluidic device. Check for any leaks by simply flushing in some PBS buffer through the inlets.

19. Prime the devices for 10–15 min before running the samples. Use syringe pumps to control the flow rates, and capture flow videos with a microscope and a high-speed camera.

Check for any air bubbles or debris that can disturb flows in the microfluidic device. The retort stands should also be at the same height for all experiments to ensure consistency in pressure.

20. Connect the loaded syringe to one of the spiral inlets via a precision syringe tip and 1.5-mm-diameter silicone tubing, and then insert separate tubing (i.e., with identical length ($\sim$15 cm)) to the outlets of the spiral microfluidic chip to deliver the output sample in target collection tubes.

21. Mount the syringe on the syringe pump and run the 70% (vol/vol) ethanol to the spiral microfluidic chip at appropriate flow rate (750 μL/min) for 5 min to sterilize the system and to get rid of bubbles within the system.

22. Load priming solution (1 × PBS and 2 mM EDTA supplemented with 0.5% (wt/vol) BSA) into a separate 60-mL syringe, run the sheath buffer through the system for 5 min to coat the surface of microchannels and wash the residual ethanol from the system.

23. Use the inverted microscope and the bright-field mode to inspect the microfluidic channels and ensure that no air bubbles are trapped in the channels. If bubbles are identified, increasing the flow rate of sheath buffer may help remove the bubbles. The chip is now ready for sample processing.

2.2.4 Blood sample preparation

24. Fresh blood samples drawn from patients must be transferred into EDTA Vacutainer tube.

Informed consent must be obtained before blood collection from human subjects. Collection must conform to all relevant governmental and institutional regulations. Further, all blood samples must be processed within 2–4 h after collection.

25. To lyse blood, warm the blood sample to room temperature (24–26 °C) and add 3 × RBC lysis buffer at a 1:3 vol/vol ratio. Invert the conical tube to mix, and incubate the mixture for 5 min at room temperature on a shaking platform or with periodic inversions, until the color changes to dark red.

26. After the incubation, collect the cells by centrifugation at 1000*g* for 5 min at room temperature, and resuspend the cell pellet in 1 × PBS to the desired concentration optimized for the protocol (2 × concentrated, $\sim$14 × 10^6 nucleated cells per mL). Mix it well to resuspend the pellet by tapping the tube gently or by mild pipetting. *Note: Lysed blood samples can be stored for 2–4 h (preferably ice) at 4 °C until processing.*

2.2.5 Blood sample processing using spiral microfluidic chip

27. As outlined in step 23 place the spiral microfluidic chip under an inverted microscope (Fig. 5) and adjust the microscope objective (10 × magnification) to see the spiral outlet.

28. Place the sterile 15-mL (for enriched sample collection) and 50-mL (for waste collection) conical tubes at the waste and sample outlet collection point.

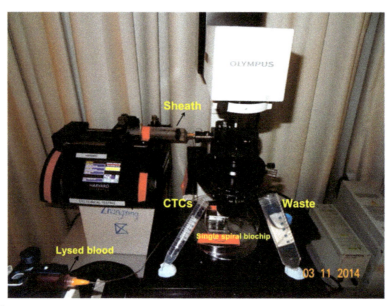

FIG. 5

Workstation setup for CTC separation. The lysed blood and one sheath flow are pumped through the spiral microfluidic chip using two different syringe pumps where CTCs are separated from other blood components rapidly and efficiently.

Images were adapted and reproduced with permission from Warkiani, M.E., Khoo, B.L., Wu, L., Tay, A.K.P., Bhagat, A.A.S., Han, J., et al. (2016). Ultra-fast, label-free isolation of circulating tumor cells from blood using spiral microfluidics, Nature Protocols, 11, 134–148 (2016). Copyright 2016 Nature Publishing Group.

29. Connect the 50-mL syringe filled with running buffer to the sheath inlet of the spiral microfluidic chip. The flow rate of this pump should be set to 750 μL/min for processing.

30. Prime the biochip by running the sheath buffer (1 × sterile PBS with 0.5% (wt/vol) BSA) at the present flow rate of 750 μL/min for ∼2 min.

31. Load the sample syringe containing lysed sample into the second pump, and then adjust the flow rate to 100 μL/min. Connect the syringe to the chip using Tygon tubing and a precision tip. *Note: Extra care should be taken for this step to ensure that the syringe plunge is not being pushed excessively and spilling the sample.*

32. Prior to sample processing, flow in the sheath buffer for ∼1 min, until the flow stabilizes. After stabilization, guide the sample collection tubing into the 15-mL conical tube, start the sample pump and continue the collection for ∼10–30 min (for 7.5 mL of lysed blood).

33. Transfer the sample outlet tube to the single cell capture chip for obtaining individual CTCs. For single CTC isolation go to step 46. For, immunofluorescence staining of isolated CTCs, imaging and counting follow steps 34–45.

Note: For isolation experiments, from spiked cancer cell lines, spike in the desired number of cancer cells before RBC lysis in step 25. To create a blood sample with a low number of cancer cells, dilute the cell suspension to $\sim 1 \times 10^5$ cells per mL. Add 10 μL of the diluted cell stock to a hemocytometer slide and image it. Manually count the cells to obtain an accurate count, and then transfer the corresponding volume from the diluted cell stock to blood.

2.2.6 Immunofluorescence staining on isolated cells

34. Concentrate the samples collected in step 33 by centrifuging at $300g$ for 3 min at room temperature.

35. Remove excess supernatant and resuspend the cells in ~ 250 μL of PBS.

36. Fix the cells with 4% (wt/vol) paraformaldehyde for 10 min at room temperature.

37. Wash the fixed cells with 1–2 mL of PBS buffer supplemented with 0.5% (wt/vol) BSA, and then centrifuge the mixture at $300g$ for 3 min at room temperature. *Note: If required, fixed cells can be stored overnight in PBS buffer at 4 °C for further use.*

38. Remove excess supernatant and resuspend the cells in ~ 250 μL containing 0.1% (vol/vol) Triton X-100 to permeabilize the cells for 1–5 min at room temperature.

39. Wash the permeabilized cells with 1–2 mL of PBS buffer supplemented with 0.5% (wt/vol) BSA, and centrifuge it at $300g$ for 3 min at room temperature.

40. Add the required conjugated antibodies directly to the cell suspension, and then incubate it on ice for 30 min. Ideally, antibodies such as FITC-conjugated pan-cytokeratin (CK) antibody (1:100) and APC-conjugated CD45 antibody (1:100) for identifying putative CTCs. However, this could depend on the nature of the sample and specific cancer type of the patient. Further, add Hoechst dye (1 M, 1:1000) to the staining solution for nuclei staining.

41. Wash stained cells with 1–2 mL of PBS buffer supplemented with 0.5% (wt/vol) BSA, and then centrifuge the mixture at $300g$ for 3 min at room temperature.

42. Resuspend the stained cells into ~ 250 μL PBS and transfer them to a single well of a 96-well plate.

2.2.7 Counting or characterization of cells

43. Count the cells using a custom-made imaging platform or manually.

44. For imaging using a custom-made platform, image and scan each well. Image the plates, using an inverted microscope (emission filters ET460/50m, ET535/50m and ET 605/70), which has an automated stage. Each well is scanned in a 1 mm × 1 mm grid format using the Metamorph software. (In our case we used an Olympus inverted microscope.) For thorough mapping and enumeration of cells in a single well, select the manual threshold function in the Image J software to select for particles corresponding to desired selection.

Table 1 Troubleshooting Problems With Spiral Chip

Step	Problem	Possible Cause	Solution
17	Incorrect particle and cell focusing	Incorrect channel height (i.e., it must be 165–170 µm), which can be the result of improper DRIE	A new master mold must be fabricated
18	Residual PDMS is trapped in the spiral channel	PDMS is trapped during the punching process	Increase the flow rate of the sheath buffer or replace the chip with a new one
22	Air bubbles are trapped in the spiral channel	Bubbles in the buffer or sample	Increase the flow rate of the sheath buffer or replace the chip with a new one
32	No stabilization in the outlet	There is a bubble trapped in the middle of channels or incorrect flow rate	Make sure that the entire chip is bubble-free and that all the parameters (flow rate and syringe diameter) are correct
32	Clogging	Large clusters are trapped in the bifurcation point	Stop the enrichment process, and clean the chip by increasing the flow rate of the sheath flow or by replacing with a new chip
45	False positives	Insufficient blocking, expired antibody reagents, inappropriate antibody concentration	Increase BSA concentration to at least 2%, and reduce the antibody concentration to 1:150. Use fresh antibodies
45	No signals detected	Insufficient permeabilization or inappropriate fixation	Fresh acetic acid/methanol fixative should be prepared for use instead of PFA

45. In case of manual imaging, obtain images of cells at $40\times$ magnification. Compare the corresponding image sets. Identify Hoechst-positive/pan-CK+/CD45-negative (CD45−) cells with round nuclei and high nuclear-to-cytoplasmic (N/C) ratio. These cells are considered to be putative CTCs and their ratio compared with other cell types is determined.

Some troubleshooting advice on potential problems that might arise during samples processing is summarized in Table 1.

3 ISOLATION AND RETRIEVAL OF SINGLE CTCs

The single cell capture chip operates based on the efficient trapping of single cells in suspension close to the curved region (Fig. 1B), ensuring that the trap chambers are in close proximity to cells and facilitate capture. A viscous flow buffer is used in a crossflow channel so as to focus the cell flow stream to the outer side (Fig. 6A).

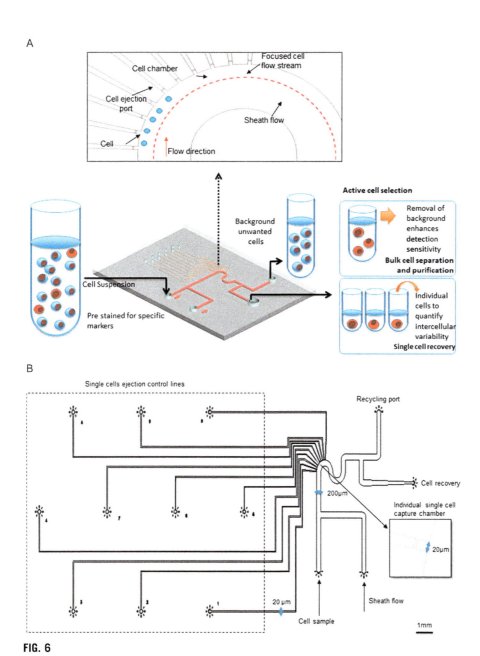

FIG. 6

Single cell capture chip setup. Schematic illustration of (A) single cell capture chip setup with inset representing cell isolation chambers, and (B) key design parameters of the chip.

Image adapted and reproduced with permission from Yeo, T., Tan, S. J., Lim, C. L., Lau, D. P. X., Chua, Y. W., Krisna, S. S., et al. (2016). Microfluidic enrichment for the single cell analysis of circulating tumor cells. Scientific Reports, 6, 22076. https://doi.org/10.1038/srep22076. https://www.nature.com/articles/srep22076#supplementary-information. Copyright Nature Publishing Group.

The use of a higher viscosity liquid allows operation under low flowrates and also minimize volume of the buffer while achieving optimal cell flow width. Ideally, the dimension of the cell flow stream should match the size of a typical cell to ensure the cells flow in a single file near the capture chambers. Considering this aspect, we achieved a 20–25 µm cell flow width. The device design, printed by soft lithography and fabricated using PDMS consist of 10 capture channels. The mechanism of aligning cells toward the channel walls makes use of two superposed fluids with different viscosities providing laminar flow. Basically, cells focused by a higher viscosity liquid lined up at the surface and gets drawn into the 1 of the 10 traps sequentially until the traps are filled. Individual cells are ejected out and selected based on morphology or positive selection with an antibody marker. Target cells that are not caught during the first run would be collected in an outlet. This allows reruns of the same sample infinite times to ensure thorough scanning of a heterogeneous sample.

3.1 MATERIALS
- PDMS and curing agent (SYLGARD® 184)
- Glass slide
- 0.5 mm and 2.0 mm biopsy puncher
- 100% Isopropanol (IPA)
- Magic Tape
- Falcon Tube
- Scapel
- Sheath Buffer Syringe 20 mL (Terumo)
- Sample Syringe 1 mL (Terumo)
- Output Microtubes 1.5 mL (Axygen®)
- Tubing
- L-pins
- Tubing clamps
- Glycerol sheath buffer (65%; ThermoFisher Scientific)
- Priming buffer (PBS 1 ×) (Axil Scientific)
- FITC-conjugated CD45 antibody (1:100; Miltenyi Biotec Asia Pacific)
- Hoechst dye (Life Technologies, cat. no. 33342)

3.2 PROCEDURE
3.2.1 Fabrication of single cell capture chips
Note: The mold fabrication was performed as described previously in Section 2.1

46. Mix PDMS base with PDMS curing agent homogeneously at a weight ratio of 10:1. With the silanized wafer master secured (using a double-sided adhesive) in the center of a 15-cm-diameter Petri dish, pour 77 g of the PDMS mixture into the dish.

47. Degas the PDMS in a vacuum desiccator for ~30 min. When the PDMS is bubble-free, bake the dish with master at 70–80 °C inside the oven for at least 2 h to cure the PDMS.

48. Cut and peel the cured PDMS from the master.

49. Punch holes to create inlets (0.5 mm dia.) and control outlets of the device at appropriate points for each PDMS piece. Punch a 2 mm dia. hole at the collection outlet.

50. Clean all the PDMS surfaces by flushing chips with ample 100% IPA and dry with an air gun. Further, clean the surface again with Scotch tape by gentle tapping. Ensure that the entire area is in contact with the tape, and check for any dirt visually.

51. Immediately, plasma-clean the PDMS pieces with features and the glass slide with surfaces to be bonded facing up for a 3 min at 300 W RF power and $760-10^{-3}$ Torr pressure.

52. Bond the featured PDMS piece to the glass slide by bringing the bonding faces into contact, and press the device softly for 30 s with a pair of tweezers to complete the bonding. *Note: To avoid channel collapse, do not apply pressure on top of the microchannel feature directly.*

53. Place the bonded PDMS device on an 85 °C hot plate or oven for 30 min to further strengthen the bonding. Allow cooling for 5 min and store the chips in a clean environment before further use.

3.2.2 Chip setup and priming

54. Connect tubing to the inlet and outlets of the microfluidic device with the ends attached to an L-shaped pin connecting the PDMS surface using the following parameters and illustration (Fig. 6B) below:

Port	Tubing Length and Diameter (in.)
Waste outlet	3.14/0.02 (ID)
Sheath outlet	1.65/0.02 (ID)
Control ports	6.10/0.02(ID)
Collection	1.88/0.04(ID)

55. Prime device with 1 × PBS from the *cell* flow port (Fig. 7A). Ensure no air bubbles are visible in the *control* tubings and secure all tubing edges consecutively with pinch clamp (i.e., inlets 1–10).

3.2.3 Sample preparation

Note: Sample obtained from spiral microfluidic chip processing (step 33) is utilized to separate individual CTCs. White blood cells and cancer cells are stained with CD45 antibody and Hoechst to allow identification and depletion of the WBCs prior to processing. Cancer cells will be visible as Hoechst + and CD45 −

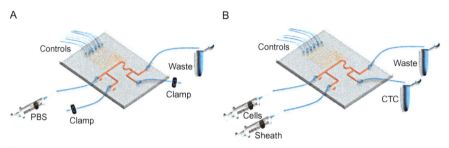

FIG. 7

Priming and sample processing using single cell capture chip. Schematic illustration of (A) chip priming using sample buffer, and (B) sample processing using a high viscous sheath buffer to enable hydrodynamic focusing of single cells into individual chambers.

56. Spin sample enriched from the spiral microfluidic chip system at 450*g* for 10 min.
57. Remove supernatant and add blocking buffer. Incubate for 15 min.
58. Spin the sample down at 450*g* for 5 min.
59. Remove supernatant, add 90 μL of blocking buffer and add 10 μL of CD45 antibody. Incubate for 30 min in dark.
60. Add Hoechst stain to the sample and incubate for 5 min.
61. Wash the cells by removing the supernatant and add 1 mL of 1 × PBS.

3.2.4 Sample processing using single cell capture chip

62. Replace the priming syringe with the cell flow setup (Fig. 7B). Transfer cell suspension to 1 mL syringe.
63. Replace sheath flow tubing with sheath flow setup. Use filtered 65% Glycerol or 18.85% PEG as the sheath fluid connected to a 20-mL syringe. Run the sheath fluid through the system for 5 min to coat the surface of microchannels and wash any residue from the system.
64. Use the inverted microscope and secure the device to the platform using a tape. Insert an Eppendorf tube for waste collection. Position tubings for control outlets so that it will be facing user.
65. Use the bright-field mode to inspect the microfluidic channels and ensure that no air bubbles are trapped in the channels (Fig. 8A). If bubbles are identified, increasing the flow rate of sheath buffer may help remove the bubbles. The chip is now ready for sample processing.
66. **START** sheath flow and immediately release cell flow pinch clamp and **START** cell flow. Observe main channel to visualize proper separation of flows. If no separation is achieved for 30 s, stop both cell and sheath flow. Repeat step this again until a laminar flow can be observed for both solutions.

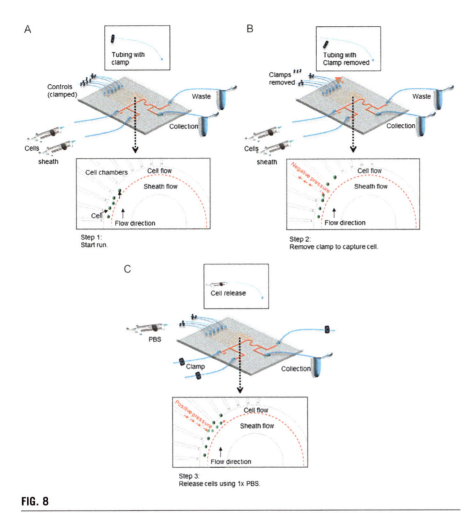

FIG. 8

Cell capture and retrieval using single cell capture chip. Schematic illustration of (A) priming of microfluidic chip, (B) focusing cells into chambers using fluid flow to trap single cells, and (C) retrieving of selected cells using positive pressure applied to specific chambers.

67. Lightly exert pressure using two fingers on Tubing 1, and observe using microscope, for slight disruption in pressure from capture channel 1 under $20\times$ magnification.

68. Allow cell flow to reach roughly 15–20 μm in size before releasing pinch clamp of tubing 1 (Fig. 8B).

69. Ensure an Eppendorf tube is positioned below tubing with pinch clamp opened to collect any exiting fluid beneath tubing one exit. If fluorescence is required, the light-source can be turned off.

Table 2 Troubleshooting Problems With Single Cell Capture Chip

Step	Problem	Possible Cause	Solution
66	Clogging	Debris observed in channel	Unplug sample and sheath. Flush device with 1 × PBS for 5 min
68	Cell flow focusing more than 20 μm	There is a bubble trapped in the middle of channels or incorrect flow rate	Stop syringe pumps. When no more flow observed in the main channel, restart syringe pump and allow flow rate to stabilize for at least 5 min
70	No cells observed/ entering capture chambers	a. Bubbles in the sample or debris obstructing flow b. Faulty syringe	a. Unplug sample and sheath. Flush device with 1 × PBS for 5 min b. Change cell sample to a new syringe
73	More than one cell entered chamber	Improper closure of capture chamber	a. Open capture chambers to the right and eject cells from the chamber with more than one cell to allow cells to flow into the next chambers sequentially b. Eject cells and recycle sample

70. Once a cell has been positively identified, immediately PAUSE the cell flow followed by sheath flow.

71. Allow channels to equilibrate for at least 3 min before securing pinch-clamp on sheath flow, followed by cell flow.

72. Carefully remove tubing from collection port, and insert a push pin gradually into the waste tubing end. Allow to equilibrate for 1 min.

73. Attached 1 × PBS syringe to tubing 1 and gradually increase force on syringe plunger until selected trapped cell moves to collection port (Fig. 8C).

74. Using a 10 μL pipette, aspirate cell from the collection port into a PCR tube. Store single cells at −20 °C for further use including downstream DNA analysis.

Some troubleshooting advice on potential problems that might arise during samples processing is summarized in Table 2.

4 CONCLUSION

CTC isolation and analysis are important first steps toward obtaining comprehensive understanding on the role of these cells in cancer metastasis and realizing their potential as liquid biopsy biomarkers. In this regard, the presented methods offer several distinct advantages over several existing microfluidic methods for CTC isolation which include: (*i*) capability to process 7.5 mL of blood in under 45 min with over 90% CTC isolation efficiency from cancer patient samples, offering unprecedented throughput and separation performance, (*ii*) minimal shear exposure to CTCs under fluid flow (<10 ms residence time) in both spiral microfluidic and single cell capture chip, thereby avoiding any undesirable change to CTC phenotype, (*iii*) simple device operation and fluid handling ensure minimal variation in isolation efficiency in multi-user setups. This is particularly important for successful implementation in clinical settings where numerous technicians and clinicians can simply replicate the experiments with minimal training, and (*iv*) the combination of spiral microfluidic and single cell capture chip can potentially bridge the gap in understanding CTC heterogeneity with the ability to retrieve single CTCs for downstream molecular analysis. Together, we believe these aforementioned advantages of this approach presents a versatile separation technique to retrieve bulk as well as single CTCs from blood and can potentially be integrated into clinical settings as a simple tool for cancer management.

REFERENCES

Adams, A. A., Okagbare, P. I., Feng, J., Hupert, M. L., Patterson, D., Göttert, J., et al. (2008). Highly efficient circulating tumor cell isolation from whole blood and label-free enumeration using polymer-based microfluidics with an integrated conductivity sensor. *Journal of the American Chemical Society, 130*(27), 8633–8641. https://doi.org/10.1021/ja8015022.

Alix-Panabieres, C., & Pantel, K. (2013). Circulating tumor cells: Liquid biopsy of cancer. *Clinical Chemistry, 59*(1), 110–118. https://doi.org/10.1373/clinchem.2012.194258. Epub 2012 Sep 26.

Bendall, S. C., & Nolan, G. P. (2012). From single cells to deep phenotypes in cancer. *Nature Biotechnology, 30*, 639. https://doi.org/10.1038/nbt.2283.

Bhagat, A. A. S., Kuntaegowdanahalli, S. S., & Papautsky, I. (2008). Enhanced particle filtration in straight microchannels using shear-modulated inertial migration. *Physics of Fluids, 20*(10), 101702. https://doi.org/10.1063/1.2998844.

Blainey, P. C., & Quake, S. R. (2013). Dissecting genomic diversity, one cell at a time. *Nature Methods, 11*, 19. https://doi.org/10.1038/nmeth.2783.

Cristofanilli, M., Budd, G. T., Ellis, M. J., Stopeck, A., Matera, J., Miller, M. C., et al. (2004). Circulating tumor cells, disease progression, and survival in metastatic breast cancer. *The New England Journal of Medicine, 351*(8), 781–791. https://doi.org/10.1056/NEJMoa040766.

Cristofanilli, M., Hayes, D. F., Budd, G. T., Ellis, M. J., Stopeck, A., Reuben, J. M., et al. (2005). Circulating tumor cells: A novel prognostic factor for newly diagnosed metastatic breast cancer. *Journal of Clinical Oncology, 23*(7), 1420–1430. https://doi.org/10.1200/JCO.2005.08.140.

Di Carlo, D., Irimia, D., Tompkins, R. G., & Toner, M. (2007). Continuous inertial focusing, ordering, and separation of particles in microchannels. *Proceedings of the National Academy of Sciences of the United States of America, 104*(48), 18892–18897. https://doi.org/10.1073/pnas.0704958104.

Gerlinger, M., Rowan, A. J., Horswell, S., Larkin, J., Endesfelder, D., Gronroos, E., et al. (2012). Intratumor heterogeneity and branched evolution revealed by multiregion sequencing. *New England Journal of Medicine, 366*(10), 883–892. https://doi.org/10.1056/NEJMoa1113205.

Gertler, R., Rosenberg, R., Fuehrer, K., Dahm, M., Nekarda, H., & Siewert, J. R. (2003). Detection of circulating tumor cells in blood using an optimized density gradient centrifugation. In H. Allgayer, M. M. Heiss, & F. W. Schildberg (Eds.), *Molecular staging of cancer* (pp. 149–155). Berlin, Heidelberg: Springer Berlin Heidelberg.

Gleghorn, J. P., Pratt, E. D., Denning, D., Liu, H., Bander, N. H., Tagawa, S. T., et al. (2010). Capture of circulating tumor cells from whole blood of prostate cancer patients using geometrically enhanced differential immunocapture (GEDI) and a prostate-specific antibody. *Lab on a Chip*, *10*(1), 27–29. https://doi.org/10.1039/B917959C.

Greaves, M., & Maley, C. C. (2012). Clonal evolution in cancer. *Nature*, *481*, 306. https://doi.org/10.1038/nature10762. https://www.nature.com/articles/nature10762#supplementary-information.

Hayes, D. F., Cristofanilli, M., Budd, G. T., Ellis, M. J., Stopeck, A., Miller, M. C., et al. (2006). Circulating tumor cells at each follow-up time point during therapy of metastatic breast cancer patients predict progression-free and overall survival. *Clinical Cancer Research*, *12*(14), 4218.

Hou, H. W., Warkiani, M. E., Khoo, B. L., Li, Z. R., Soo, R. A., Tan, D. S.-W., et al. (2013). Isolation and retrieval of circulating tumor cells using centrifugal forces. *Scientific Reports*, *3*, 1259. https://doi.org/10.1038/srep01259. https://www.nature.com/articles/srep01259#supplementary-information.

Huang, S.-B., Wu, M.-H., Lin, Y.-H., Hsieh, C.-H., Yang, C.-L., Lin, H.-C., et al. (2013). High-purity and label-free isolation of circulating tumor cells (CTCs) in a microfluidic platform by using optically-induced-dielectrophoretic (ODEP) force. *Lab on a Chip*, *13*(7), 1371–1383. https://doi.org/10.1039/C3LC41256C.

Hur, S. C., Henderson-MacLennan, N. K., McCabe, E. R. B., & Di Carlo, D. (2011). Deformability-based cell classification and enrichment using inertial microfluidics. *Lab on a Chip*, *11*(5), 912–920. https://doi.org/10.1039/C0LC00595A.

Hyun, K.-A., Kwon, K., Han, H., Kim, S.-I., & Jung, H.-I. (2013). Microfluidic flow fractionation device for label-free isolation of circulating tumor cells (CTCs) from breast cancer patients. *Biosensors and Bioelectronics*, *40*(1), 206–212. https://doi.org/10.1016/j.bios.2012.07.021.

Klein, C. A. (2013). Selection and adaptation during metastatic cancer progression. *Nature*, *501*, 365. https://doi.org/10.1038/nature12628.

Kuntaegowdanahalli, S. S., Bhagat, A. A. S., Kumar, G., & Papautsky, I. (2009). Inertial microfluidics for continuous particle separation in spiral microchannels. *Lab on a Chip*, *9*(20), 2973–2980. https://doi.org/10.1039/B908271A.

Luo, J., Solimini, N. L., & Elledge, S. J. (2009). Principles of cancer therapy: Oncogene and non-oncogene addiction. *Cell*, *136*(5), 823–837. https://doi.org/10.1016/j.cell.2009.02.024.

Nagrath, S., Sequist, L. V., Maheswaran, S., Bell, D. W., Irimia, D., Ulkus, L., et al. (2007). Isolation of rare circulating tumour cells in cancer patients by microchip technology. *Nature*, *450*, 1235. https://doi.org/10.1038/nature06385. https://www.nature.com/articles/nature06385#supplementary-information.

Pantel, K., Brakenhoff, R. H., & Brandt, B. (2008). Detection, clinical relevance and specific biological properties of disseminating tumour cells. *Nature Reviews Cancer*, *8*, 329. https://doi.org/10.1038/nrc2375.

Pecot, C. V., Bischoff, F. Z., Mayer, J. A., Wong, K. L., Pham, T., Bottsford-Miller, J., et al. (2011). A novel platform for detection of CK + and CK − CTCs. *Cancer Discovery*, *1*(7), 580–586. https://doi.org/10.1158/2159-8290.cd-11-0215.

Polzer, B., Medoro, G., Pasch, S., Fontana, F., Zorzino, L., Pestka, A., et al. (2014). Molecular profiling of single circulating tumor cells with diagnostic intention. *EMBO Molecular Medicine*, *6*(11), 1371–1386. https://doi.org/10.15252/emmm.201404033.

Powell, A. A., Talasaz, A. H., Zhang, H., Coram, M. A., Reddy, A., Deng, G., et al. (2012). Single cell profiling of circulating tumor cells: Transcriptional heterogeneity and diversity from breast cancer cell lines. *PLoS One, 7*(5). e33788. https://doi.org/10.1371/journal.pone.0033788.

Saliba, A.-E., Saias, L., Psychari, E., Minc, N., Simon, D., Bidard, F.-C., et al. (2010). Microfluidic sorting and multimodal typing of cancer cells in self-assembled magnetic arrays. *Proceedings of the National Academy of Sciences of the United States of America, 107*(33), 14524–14529. https://doi.org/10.1073/pnas.1001515107.

Spizzo, G., Gastl, G., Obrist, P., Went, P., Dirnhofer, S., Bischoff, S., et al. (2004). High Ep-CAM expression is associated with poor prognosis in node-positive breast cancer. *Breast Cancer Research and Treatment*, *86*(3), 207–213. https://doi.org/10.1023/B:BREA.0000036787.59816.01.

Stott, S. L., Hsu, C.-H., Tsukrov, D. I., Yu, M., Miyamoto, D. T., Waltman, B. A., et al. (2010). Isolation of circulating tumor cells using a microvortex-generating herringbone-chip. *Proceedings of the National Academy of Sciences*, *107*(43), 18392–18397. https://doi.org/10.1073/pnas.1012539107.

Warkiani, M. E., Guan, G., Luan, K. B., Lee, W. C., Bhagat, A. A. S., Kant Chaudhuri, P., et al. (2014). Slanted spiral microfluidics for the ultra-fast, label-free isolation of circulating tumor cells. *Lab on a Chip*, *14*(1), 128–137. https://doi.org/10.1039/C3LC50617G.

Warkiani, M. E., Khoo, B. L., Wu, L., Tay, A. K. P., Bhagat, A. A. S., Han, J., et al. (2016). Ultra-fast, label-free isolation of circulating tumor cells from blood using spiral microfluidics. *Nature Protocols*, *11*, 134. https://doi.org/10.1038/nprot.2016.003. https://www.nature.com/articles/nprot.2016.003#supplementary-information.

Yeo, T., Tan, S. J., Lim, C. L., Lau, D. P. X., Chua, Y. W., Krisna, S. S., et al. (2016). Microfluidic enrichment for the single cell analysis of circulating tumor cells. *Scientific Reports*, *6*, 22076. https://doi.org/10.1038/srep22076. https://www.nature.com/articles/srep22076#supplementary-information.

High-throughput single-cell mechanical phenotyping with real-time deformability cytometry

10

Marta Urbanska*, Philipp Rosendahl*, Martin Kräter*, Jochen Guck[1]

Biotechnology Center, Center for Molecular and Cellular Bioengineering,
Technische Universität Dresden, Dresden, Germany
[1]Corresponding author: e-mail address: jochen.guck@tu-dresden.de

CHAPTER OUTLINE

---

*These authors contributed equally to this work.

Methods in Cell Biology, Volume 147, ISSN 0091-679X, https://doi.org/10.1016/bs.mcb.2018.06.009

Abstract

Mechanical properties of cells can serve as a label-free marker of cell state and function and their alterations have been implicated in processes such as cancer metastasis, leukocyte activation, or stem cell differentiation. Over recent years, new techniques for single-cell mechanical characterization at high throughput have been developed to accelerate discovery in the field of mechanical phenotyping. One such technique is real-time deformability cytometry (RT-DC), a robust technology based on microfluidics that performs continuous mechanical characterization of cells in a contactless manner at rates of up to 1000 cells per second. This tremendous throughput allows for comparison of large sample numbers and precise characterization of heterogeneous cell populations. Additionally, parameters acquired in RT-DC measurements can be used to determine the apparent Young's modulus of individual cells. In this chapter, we present practical aspects important for the implementation of the RT-DC methodology, including a description of the setup, operation principles, and experimental protocols. In the latter, we describe a variety of preparation procedures for samples originating from different sources including 2D and 3D cell cultures as well as blood and tissue-derived primary cells, and discuss obstacles that may arise during their measurements. Finally, we provide insights into standard data analysis procedures and discuss the method's performance in light of other available techniques.

1 INTRODUCTION

Mechanical properties of cells reflect their state and function and, thus, can be used as an intrinsic biophysical marker for the detection of pathological cell changes or the determination of physiological cell state transitions (Di Carlo, 2012; Nematbakhsh & Lim, 2015). Several techniques have been introduced over the past decades to examine cell deformation under exposure to external stress. More established methods include atomic force microscopy (AFM) (Radmacher, 2007), micropipette aspiration (Hochmuth, 2000), and optical stretcher (Guck et al., 2001). In recent years, advancements in microfabrication technology have brought about the development of microfluidic-based techniques for mechanical characterization of cells. Whether they rely on measuring the cell transit time through a microconstriction (Byun et al., 2013; Lange et al., 2015) or on estimating cell deformation caused by hydrodynamic forces (Gossett et al., 2012; Nyberg et al., 2017), they are all able to perform at unprecedented throughput, enabling thousands of single-cell recordings within minutes.

One such microfluidics-based technique is real-time deformability cytometry (RT-DC) (Otto et al., 2015), a method recently developed in our laboratory. Based on the hydrodynamic deformation of cells translocating in a contact-free manner through a microfluidic channel, RT-DC is able to continuously analyze up to 1000 cells per second in real time. With such a measurement speed, comparing mechanical properties of different cell populations or finding mechanical subpopulations within measured samples is greatly facilitated, and the real-time image processing allows for instantaneous evaluation of obtained results. Furthermore, an analytical model (Mietke et al., 2015) and numerical simulations (Mokbel et al., 2017) of RT-DC measurements have been developed to enable assigning an apparent Young's modulus value to each cell to parameterize cell stiffness.

In this chapter, we provide a detailed description of the setup and experimental protocols used to perform mechanical characterization of single cells with RT-DC. We present a variety of preparation protocols for samples originating from different sources, discuss obstacles that may arise during their measurements, and provide insights into standard data analysis procedures.

2 METHOD OVERVIEW
2.1 EXPERIMENTAL SETUP

An RT-DC setup is built around a commercially available inverted microscope (such as Axiovert 200M, Zeiss, Germany). It consists of a high-precision syringe pump (two modules neMESyS 290N, NeMESyS, Cetoni, Germany), a high-speed CMOS camera (MC1365, Mikrotron, Germany), an LED lamp for stroboscopic illumination that is synchronized with the camera (AcCellerator L1, Zellmechanik Dresden, Germany), and a stage with a customized holder on which a microfluidic chip is installed during the measurements (Fig. 1A). The whole setup is controlled by a user interface co-developed with Zellmechanik Dresden (*ShapeIn* and *PumpOperation*) on a standard personal computer.

2.2 METHOD RATIONALE

RT-DC relies on flowing cells through a constriction in a microfluidic channel in which they are deformed in a contactless manner by hydrodynamic shear and normal stresses (Fig. 1B). The extent of deformation of the cell is evaluated based on high-speed optical imaging. A custom algorithm detects the cell contour and quantifies its deformation, defined as the deviation from a circular cross-section as described by the equation in Fig. 1B, and size, defined as the cross-sectional area of the contour, in real time. The results are then displayed in a size vs. deformation scatter plot (Fig. 1C). Since larger cells experience higher stresses in the constriction, the deformation is not a direct measure of cell deformability. A similarly stiff, but larger cell necessarily deforms more. To decouple size and deformation, an analytical model

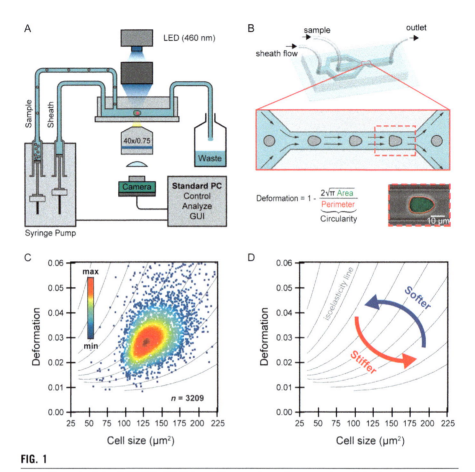

FIG. 1

Operation principle of RT-DC. (A) Schematic overview of an experimental setup consisting of a syringe pump, an LED light source, and a camera installed on an inverted microscope. (B) Cell deformation in the channel constriction. The upper diagram shows a 3D representation of the microfluidic chip with inlets for sheath and sample flows and an outlet. The close-up shows the channel constriction with the region of interest for image processing indicated by the red dashed box. The inset presents an image of a deformed cell. Deformation is quantified as the deviation from a circular cross-section as defined by the equation. (C) Typical results of an RT-DC measurement presented on a deformation vs. cell size scatter plot. Each dot represents the result of a single cell measured out of a total of $n=3209$ cells, the color scale indicates event density. (D) Isoelasticity lines derived from numerical simulations aid in identifying cells of corresponding mechanical properties in a deformation vs. cell size plot.

(Mietke et al., 2015) as well as numerical simulations (Mokbel et al., 2017) have been developed to provide reference isoelasticity lines for finding cells of different sizes with corresponding stiffness (Fig. 1D) and to extract an apparent Young's modulus for each cell, under assumption that the cell is a homogeneous, isotropic sphere.

2.3 OPERATION DETAILS

As depicted in Fig. 1A, a constant flow of medium and cell suspension (0.01–1 μL/s) is created by a computer-controlled syringe pump. In the microfluidic chip (Figs. 1B and 2), the cell suspension is led into a constriction zone centered by sheath flows from the second syringe. The ratio between sample and sheath flow rate is 1:3, which has proven to deliver best results for cell focusing. In the constriction, which has a square cross-section with typical widths ranging from 15 to 40 μm, the suspended cells are subject to shear and normal stresses caused by the flow velocity being higher in the center of the channel as compared to regions closer to the channel walls. These stresses lead to characteristic "bullet-like" cell shapes in the steady state at the end of the channel constriction as depicted in Fig. 1B. To quantify the deformation, the microfluidic chip is located on an inverted microscope equipped with a high-speed CMOS camera and a synchronized stroboscopic LED bright-field illumination. Stroboscopic illumination is necessary because cells in a 20 μm channel with a typical flow rate of 0.04 μL/s have a velocity of about 0.15 m/s. To reduce motion blurring to less than 0.5 μm (resolution of the microscope), the exposure time t_e needs to be:

$$t_e < \frac{0.5\,\mu m}{0.15\,m/s} = 3.3\,\mu s.$$

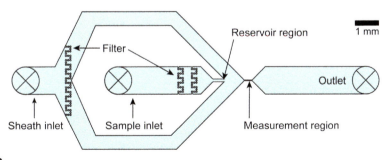

FIG. 2

Layout of the microfluidic chip for RT-DC experiments. Tubings with sheath and sample fluid, as well as outlet tubing are connected to the holes punched through the PDMS (marked with crosses). After passing micro-pillar filters, the sheath and sample streams are combined. In the narrow channel of the measurement region, hydrodynamic stresses deform the cells. After the measurement, the fluid leaves the chip through the waste outlet on the right.

The intensity of common halogen tungsten microscope illuminations is not sufficient for such small exposure times. We found that a lamp using a high-power LED (CBT-90, Luminus Devices, CA, USA) with a custom driver circuit provides the necessary intensities with an adjustable pulse length between 1 and 10 μs. The lamp is connected to the trigger output of the high-speed CMOS camera for synchronized operation. The imagining is typically performed with a 40× objective (EC-Plan-Neofluar, 40×/0.75, Zeiss, Germany). The resulting images with negligible motion blurring are directly processed by a personal computer (PC) with a custom-written C++ software. The image processing consists of the following steps: (i) background subtraction, (ii) threshold filtering, (iii) contour finding, and (iv) contour processing for the estimation of cell size, position, deformation, and brightness, among others. For mechanical characterization, the parameter deformation is of primary interest. Deformation, d, is based on a feature called circularity, c, and is defined as:

$$d = 1 - c = 1 - \frac{2\sqrt{\pi A}}{P},$$

where A is the area enclosed by the contour and P the perimeter of the contour. Extracted parameters, the microscope images, and time stamps are stored for each single cell detected in a ".tdms" file for later analysis.

3 EXPERIMENTAL PROTOCOLS
3.1 SAMPLE PREPARATION

For RT-DC measurements, cells are suspended in a viscosity-adjusted measurement buffer (MB, final viscosity 15 mPa s for standard measurements and 25 mPa s for blood and bone marrow measurements) containing methylcellulose (for details regarding MB preparation see Section 3.2.1). The elevated buffer viscosity allows for appreciable cell deformations at moderate flow rates. At the same time, the increased density of the buffer prevents cells from sedimentation which enables continuous measurements (up to several hours) and reduces bias towards slower sedimenting cells when analyzing heterogeneous samples.

Optimal conditions for RT-DC measurements are typically obtained by adjusting the cell suspension in MB to a final concentration of $3–5 \times 10^6$ cells/mL. Theoretically, no limitation in minimal sample volume exist; however, to allow effective filling of the device a sample volume of minimum 20 μL is recommended. For convenient handling, and measurement at several flow rates, of samples with concentrations in the range given above, a total cell suspension volume of about 100 μL is sufficient.

In principle, there are no restrictions regarding the origin and type of cells subjected to RT-DC measurements; however, some biological and technical aspects need to be considered before running the measurements. As RT-DC experiments require a single-cell suspension, cell detachment from culture surface for adherent cells, or cell dissociation for 3D cell cultures and primary tissues is required. The transition from cell–cell or cell–surface contact to a single-cell state may lead to mechanical adaptation of the cells associated with F-actin filament remodeling (Maloney et al., 2010). This adaptation affects some cell types more than others and should be considered or tested for prior to performing experiments of interest. Potential mechanical changes following detachment or dissociation over time can be monitored using RT-DC by measuring at different time points after cell preparation.

Cell preparation is a crucial step for RT-DC experiments. Depending on the cell type and growth conditions, optimal sample preparation may vary. In the following, sample preparations for some classical cell culture methods and cell sources are presented.

3.1.1 Non-adherent cells

For semi-adherent cells or cells growing in suspension, simply collect the desired volume of the cell culture by pipetting, and perform the following steps:

(i) Centrifuge the cells according to standard procedures (e.g., 5 min at $150 \times g$).
(ii) Discard supernatant.
(iii) Resuspend the cell pellet thoroughly by pipetting up and down several times in 20–1000 µL of MB. Aim for a cell concentration of $3–5 \times 10^6$ cells/mL.

Notes
- Before resuspension in MB, the supernatant should be carefully and fully removed to prevent MB dilution.
- 20 µL is the minimal sample volume required for filling the channel and running a measurement at one flow rate; 1000 µL is the maximal sample volume that can be aspirated into the 1 mL syringe typically used.
- If the cell concentration in the collected culture is not known, it is useful to count the cells before spinning them down to estimate the volume of MB that will be used for resuspension.
- If necessary, the cell pellet can be washed in PBS prior to resuspension in MB to remove cell debris and prevent channel clogging in the RT-DC experiment.
- If cells are prone to cluster formation, a filtration of the cell suspension through a strainer with a mesh size of 40 µm (EASYstrainer™ #542040; Greiner Bio-One, Germany; #431750, Corning Inc., NY, USA; FlowMi #H136800040, Belart, NJ, USA or alike) can be implemented during additional washing step or after the final cell resuspension in MB.

3.1.2 Adherent cells

In case of adherent cultures, detach the cells from the dish surface to generate a single-cell suspension according to a standard protocol for the cell line used. Typically, a PBS washing step, followed by an incubation with a dissociation agent (e.g., trypsin, accutase, collagenase, or EDTA) is performed. After cells are dissociated, suspend them in a buffer quenching the activity of the dissociation agent (e.g., serum-containing buffer in case of trypsin) or simply diluting it. Subsequently, follow the procedure described for the non-adherent cells (see Section 3.1.1).

3.1.3 3D cell cultures
- Sphere cultures

Sphere-cultured cells (forming 3D clusters of cells of the same type) need to be dissociated to generate a single-cell suspension prior to the measurement. For dissociation use a protocol suitable for the system used. An exemplary protocol for non-adherent sphere culture includes the following steps:

(i) Dissociate the spheres enzymatically by incubating the suspension in medium without serum supplemented with type I collagenase, and 2.5% type I DNase. Incubation should be performed in a humidified incubator at $37\,°C$ with 5% CO_2 for up to 3 h adjusted based on the dissociation progress.
(ii) Stop dissociation by adding 25% serum.
(iii) Collect cells by centrifugation.
(iv) Wash cells with PBS and filter the cell suspension through a 40 µm cell strainer to remove cell clusters.

For further preparation, the cell suspension can be treated as described under non-adherent cells (see Section 3.1.1).

- Hydrogel

To measure cells cultured inside 3D hydrogels (such as Matrigel®, Corning Inc., NY, USA), separation from the gel and generation of a single-cell suspension is required prior to the measurement. Standard or custom protocols can be used to dissociate the hydrogel. Cell aggregates can be dissociated according to the protocol for spheres. Cells are collected by centrifugation, followed by a washing and filtration step using PBS and a 40 µm cell strainer in order to remove cell culture debris and to prevent channel clogging in the RT-DC experiment. For further preparation, the cell suspension can be treated as described under non-adherent cells (see Section 3.1.1).

- Organoids

To measure organoid-derived cells, organoids need to be dissociated in order to generate a single-cell suspension. All standard or custom protocols can be used, or dissociation can be performed as suggested in the following. For hydrogel-embedded organoids, the hydrogel needs to be dissociated as described in the hydrogel section. The obtained cell suspension can then be prepared for RT-DC measurement as described under non-adherent cells (see Section 3.1.1). Non-hydrogel embedded organoids and cell aggregates can be enzymatically dissociated as described for spheres.

3.1.4 Blood and tissue-derived primary cells

• Blood

For whole-blood RT-DC measurements, 50 μL of anti-coagulated blood is diluted in 950 μL MB and mixed gently by manual rotation of the sample tube as described in detail elsewhere (Toepfner et al., 2018). Depending on the cell population of interest, the ratio of blood to MB may be adjusted. For example, for effective measurements of the very abundant red blood cells, it is sufficient to dilute 5 μL blood in 995 μL MB.

• Bone marrow

For measurement of whole bone marrow, 50 μL of anti-coagulated bone marrow is diluted in 950 μL MB and mixed gently by manual rotation of the sample tube.

Note: For performing measurements on blood and bone marrow-derived samples, MB with viscosity of 25 mPa s is typically used.

• Solid tissues

Solid tissues need to be dissociated before RT-DC analysis, as measurements can only be performed with a single-cell suspension. As different tissues need different dissociation protocols, an appropriate protocol should be chosen for the cells of interest. In doubt, a simple protocol as suggested for spheres could be used. Once a single-cell suspension is prepared, cells can be treated as described for non-adherent cells (Section 3.1.1).

3.2 SETUP PREPARATION

The RT-DC setup needs to be prepared by mounting a microfluidic chip with selected channel width onto the microscope stage, installing the sheath and the sample syringes on the syringe pump, and connecting the chip to sheath and sample tubing.

3.2.1 Materials

A list of materials required for setting up an RT-DC measurement is provided in Table 1. In the sections below the microfluidic chips as well as the measurement buffers used for the RT-DC experiments are described in greater detail.

• Microfluidic chips

The microfluidic chip constitutes the central part of RT-DC as depicted in Fig. 1. The chip is made from polydimethylsiloxane (PDMS, SYLGARD, 188 Dow, Corning Inc., NY, USA) using soft lithography. A mixture of PDMS and curing agent (10:1, w/w) is poured over a silicon wafer master and cured as described elsewhere (Herbig et al., 2017). The holes for sheath and sample in- and outlets are punched through the PDMS replica (see Fig. 2) with the channel imprint using a 1.5-mm puncher (e.g., Biopsy Punch #49115, Pfm Medical AG, Germany). The PDMS replica are then covalently bound to a microscopy-suited cover glass ($40 \times 24\,\text{mm}^2$, Assistent, Germany) and sealed by plasma activation (50 W, 30 s, Plasma Cleaner Atto, Diener Electronic, Germany). In Fig. 2, the layout of a conventional RT-DC chip is presented. The sheath flow and the sample flow are introduced to the channel through two separate inlets. Each has a filter structure in the flow path that retain

Table 1 Consumables Necessary for Setting Up an RT-DC Experiment

Article	Product Name, Company	Order No.
FEP tubing	FEP Tubing 1/16″ OD, 0.030″ ID, Postnova Analytics, Germany	1520XL
Syringe connector part 1	PEEK Union for 1/16″ OD Tubing, Postnova Analytics, Germany	P-702
Syringe connector part 2	F Luer to ¼-28 FB, F, Postnova Analytics, Germany	P-658
Sheath fluid/ sample syringe	BD Luer-Lok™ 1-mL syringe, BD Biosciences, NJ, USA	613-4971
Syringe for tubing cleaning	BD Disposable Luer-Lok™ tip 5-mL syringe, Henke Sass Wolf, Germany	613-2043
Syringe needle	Blunt Fill Needle 18G, BD Biosciences, NJ, USA	BDAM305180
Syringe filter unit	Millex-GV, 0.22 μm, PVDF, Merck Millipore, Germany	SLGV004SL
Microfluidic chip	Flic15/Flic20/Flix30/Flic40/FlicXX, Zellmechanik Dresden, Germany	–
Measurement buffer	CellCarrier/CellCarrierB, Zellmechanik Dresden, Germany	–

bigger particles and prevent the channel constriction from clogging. The sheath flow is split into two branches that merge with sample flow from the sides to induce hydrodynamic focusing of the cells toward the channel constriction. The channel constriction in which cells are deformed is roughly 300 μm long and has a square cross-section. The cell deformation is assessed within a 85-μm long region at the end of the constriction (see Fig. 1B). Importantly, the channel width should be selected in correspondence to the cell size. For optimal performance, cell diameters should cover 20–90% of the channel width. Commonly used channel widths are 15–40 μm and are commercially available from Zellmechanik Dresden (Flic15, Flic20, Flic30, Flic40). Alternatively, the chips can be self-produced as described in detail elsewhere (Herbig et al., 2017).

- Measurement buffer (MB)

MB is prepared by dissolving 0.5% (w/v) methylceullose (MC, 4000 cPs #36718, Alfa Aesar, Germany) in a physiological buffer, such as PBS or the cell culture medium. Prolonged mixing of about 24 h on a rotary mixer is required to allow for full dissolution of MC powder in the selected medium. After MC mixes in, the buffer is filtered through a vacuum filter unit (Stericup-GP, 0.22 μm, Merc Millipore, Germany). Next, its viscosity is examined over several temperatures ranging from 18 to 30 °C using a falling drop viscometer (HAAKE, Thermo Fisher Scientific, MA, USA) and adjusted to 15 mPa s at 23 °C. For blood measurements an MB with final viscosity of 25 mPa s, prepared by dissolving 0.6% (w/v) MC in PBS, is used.

It is good practice to test the osmolarity of MB and, if necessary, adjust it to the typical growth conditions of the cells of interest to avoid cell swelling or shrinking due to osmotic effects. In case of standard cell lines or primary cells derived from human or mouse, an MB osmolarity of 320 ± 20 mOsm is adequate. For reference, RPMI-based media have an osmolarity of approximately 300 mOsm, which corresponds to the osmolarity of human blood plasma (Silverthorn & Johnson, 2009), DMEM-based media have osmolarity around 340 mOsm, and the osmolarity of PBS solutions usually falls in the range between 285 and 325 mOsm.

MBs for conventional measurements as well as for blood measurements are commercially available for purchase (CellCarrier and CellCarrierB, Zellmechanik Dresden, Germany).

3.2.2 Preparation procedure

To prepare an RT-DC measurement, go through the following steps:

(i) Switch on the operating computer, the microscope, the LED lamp, the camera, and the syringe pump.

(ii) Launch the *PumpOperation* software (Zellmechanik Dresden) for controlling the syringe pump and the *ShapeIn* software (Zellmechanik Dresden) for viewing the current camera image and recording the measurement, or similar custom-written software.

(iii) Mount a measurement chip with a selected channel width to the dedicated holder on the microscope stage.

(iv) Clean sheath-flow and sample-flow tubings by flushing them subsequently with 3 mL ethanol and 5 mL distilled water. Blow the tubing dry with compressed nitrogen.

Note: The tubing should have a length that allows for connecting the syringe mounted on the syringe pump to the chip placed on the microscope (typically around 25 cm).

(v) Drain a few milliliters of MB through a 0.22 µm syringe filter unit to remove all contaminant particles from the solution.

(vi) Use a blunt-end needle to fill the sheath syringe with up to 1 mL of filtered MB. Remove air bubbles from the syringe. Discard the needle, connect the syringe to the tubing by using a Luer-Lok connector and push the liquid through to fill the tubing. Make sure that there are no air bubbles present in the syringe and the tubing.

Note: Before mounting the syringe in steps (vii) and (viii), adjust the pump piston position using *PumpOperation* to the liquid volume present in the corresponding syringe.

(vii) Mount the sheath syringe to the pump and connect the tubing to the sheath inlet of the measurement chip. Go to *PumpOperation* and start the sheath flow with a flow rate of 0.2 µL/s to fill the channel. Monitor the process by observing the camera image of the chip with "Overview" option in the *ShapeIn* software. The chip is entirely filled when a droplet appears at the sample inlet and at the outlet. Reduce the flow to 0.1 µL/s. The chip should be free of air bubbles.

(viii) Fill the sample syringe with ca. 0.5 mL of MB using a blunt-end needle, connect it with the tubing and push the liquid through to fill the tubing. No bubbles should be present. Mount the sample syringe to the pump.

(ix) Insert the free end of the sample tubing into the vial containing cell suspension in MB (sample preparation described in Section 3.2). With *PumpOperation* apply a negative flow rate of up to $-1\,\mu L/s$ to draw an appropriate volume of sample into the tubing (at least 10 times the cell volume you intend to measure). Make sure that no air is drawn into the tubing.

Note: If there is a big excess of cell suspension over the tubing volume, it is possible to directly fill the sample syringe with sample, for example, for blood measurement.

(x) Start the sample syringe with a flow rate of $+0.1\,\mu L/s$ and connect tubing to sample inlet.

(xi) Prepare an outlet tubing and connect one of its ends to the outlet. The other end should be placed in a waste container or a collection tube if the analyzed cells are to be reused.

Note: At this point it is important to check whether any MB leaks through the chip edges or the inlets of the measurement chip. If this is the case, it is necessary to replace the chip as the laminar flow is likely disturbed and there is loss of pressure from the system.

(xii) Before performing the measurement, create a measurement folder to which the data will be saved by using "Project Manager" in *ShapeIn*.

3.3 MEASUREMENT

3.3.1 Setting filters before the measurement

Before running a measurement, set up gating boundaries for cell size and/or aspect ratio to avoid recording superfluous data. Gating out objects of very small size is recommended to prevent recording of dirt particles or small cell debris and to help keeping track of the number of the events of interest. Gating out particles of large size and large aspect ratio is helpful for discarding cell clusters. A standard setting for a 20 μm channel is a minimum height and length of 3 μm, a maximum height of 20 μm, and a maximum length of 80 μm. The maximum aspect ratio can usually be set to 2.

3.3.2 Performing the measurement

After the microfluidic chip has been filled with sheath and sample fluids in the last step of the setup preparation, continue with the following:

(i) Adjust the flow rate according to the chip size and the MB viscosity (see Table 2) and allow the flow to equilibrate for at least 1 min.

(ii) In "Overview" mode of *ShapeIn*, find the channel constriction by moving the stage of the microscope and align the indicated region of interest with the end of the measurement channel.

Table 2 Measurement Flow Rate Recommendations for Different RT-DC Channels Widths and a MB Viscosity of 15 mPas

Channel Width	Total Flow Rate (μL/s)	Sample Flow Rate (μL/s)	Sheath Flow Rate (μL/s)
10 μm	0.004	0.001	0.003
	0.008	0.002	0.006
	0.016	0.004	0.012
20 μm	0.04	0.01	0.03
	0.08	0.02	0.06
	0.12	0.03	0.09
30 μm	0.16	0.04	0.12
	0.24	0.06	0.18
	0.32	0.08	0.24
40 μm	0.32	0.08	0.24
	0.64	0.16	0.48
	0.96	0.24	0.72

Underfocus   Slight Underfocus   In focus   Overfocus

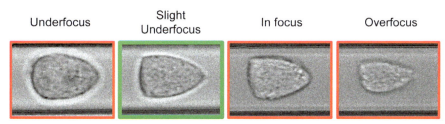

FIG. 3

Finding the right cell focus for the RT-DC measurement. For successful image thresholding, the focus of cells should be tuned to slightly under-focused. The cell should appear darker than the background with a bright halo around its contour as in the image second from the left (indicated with a green box).

Note: Check the channel and filters for potential clogging beforehand. If clogging occurs, you can try to remove it by detaching the sample tubing and transiently increasing the sheath flow rate up to 1 μL/s. If clogging persists, discard the chip and repeat the preparation procedure with a new one.

(iii) Go back to the main *ShapeIn* window and adjust the focus of cells to a slight underfocus. A thin bright halo should be visible around the cell (Fig. 3).

Note: If cells flow too close to the channel wall, in an asymmetric manner, or at varying focus, hydrodynamic focusing is disturbed. Potentially, there are air bubbles in the chip, particles clog the channel entry, the syringes are set to incorrect flow rates, or the system is not sealed.

(iv) In *ShapeIn*, check that a correct destination folder is selected. Start the measurement by pressing the "Start" button and acquire the desired number of events. For a homogeneous sample typically around 2000 cells are measured.

Note: You can stop the measurement manually or preselect the program to stop automatically after a specified time or a selected number of events is reached (in the tab "Stop Conditions").

(v) Repeat the measurement steps (i)–(iv) for the two remaining flow rates as specified for your channel width in Table 2.

Note: This is an optional step. It provides an internal control as the deformation of cells should increase with increasing flow rate.

(vi) Select the "Reservoir" measurement option in *ShapeOut*. The region of imaging will be updated to a bigger square. Position it at the 100-μm broad channel, right before the place where the sample merges with sheath fluid (see Fig. 2). Record the measurement.

Note: This is another internal control. It is a prerequisite for using the Young's modulus extraction that the cells appear spherical in the "Reservoir" measurement

(vii) Dismount the sample syringe, clean the tubing by flushing it with ethanol and water and blow-dry it with compressed nitrogen, replace the chip with a new one, and fill it with MB. Proceed with measuring the next sample.

4 DATA ANALYSIS

Analysis of RT-DC data is similar to analysis done for flow cytometry. Steps performed routinely include gating for a cell population of interest, calculation of derived parameters (as shown for Young's modulus values), and application of statistical methods such as linear mixed models for comparison of results obtained for different samples. In contrast to flow cytometry, it is possible to inspect outliers and cells of interest by looking at the actual image which is saved for every recorded event.

The following text will provide a guide through the process of data analysis. It is recommended to use the open-source software *ShapeOut* (available at https://github.com/ZELLMECHANIK-DRESDEN/ShapeOut) as it is able to read the data and settings files recorded with RT-DC and provides tools for filtering, parameter calculations, and plotting. Furthermore, *ShapeOut* has a build-in functionality to export the raw or processed data in ".csv" and ".fcs" formats. The exported data can be then used for analysis in other programs such as Excel (Microsoft, WA, USA), MatLab (MathWorks, MA, USA), Origin (OriginLab, MA, USA), or FlowJo (FlowJo LCC, OR, USA).

4.1 APPLYING FILTERS TO DATA

RT-DC performs image processing in real time during the experiment to determine the following parameters from bright-field images: cell size, deformation, area ratio, and aspect ratio. Those parameters, together with additional parameters calculated offline with *ShapeOut*, can be used for filtering and gating strategies.

Filtering of raw data is usually performed to exclude invalid events such as cell doublets or larger aggregates, cell debris, or non-intact cells, which were harmed due to handling or other reasons. Furthermore, filtering can be applied to select cell types of interest in heterogeneous samples, such as bone marrow or blood.

- Cell size

An obvious parameter for filtering is cell size, which is measured as the projected cell cross-sectional area in units of μm^2. Doublets, dead, or degenerate cells often are smaller or larger than the targeted main population. If not already done during data acquisition, small objects that usually correspond to cell debris can be excluded in post-processing. Events with very large sizes correspond likely to cell aggregates. The final values for the filters have to be determined specifically for each study, best also considering the recorded images. In *ShapeOut*, it is possible to inspect images of individual events by selecting a data point of interest on the deformation vs. cell size scatter. This valuable information helps the user in determining correct filter values.

- Area ratio

For a correct determination of cell mechanical parameters, such as deformation or the derived Young's modulus, it is important that the cell has a smooth contour. Cells with protrusions or cells that are poorly captured have an overestimated deformation. This may lead to wrong conclusions regarding their mechanical properties. To avoid this, the parameter called *area ratio* is introduced. It is defined as:

$$\text{area ratio} = \frac{A_C}{A_R},$$

where A_C is the area of the convex hull of the detected contour and A_R is the area of the initially detected contour (see Fig. 4). The more concave sections are found in the contour, the higher the *area ratio* value. Since even strongly deformed cells with smooth surface do not have concave sections in their contour, filtering for *area ratio* does not bias the measurement of deformations caused by the hydrodynamic forces. The *area ratio* is usually set to be contained between values of 1.00 and 1.05 or 1.10.

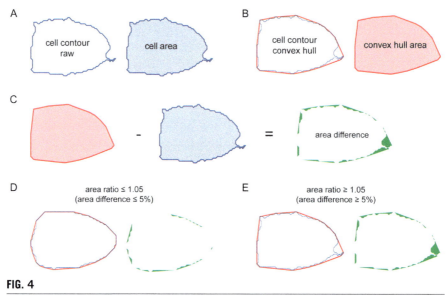

FIG. 4

Schematic representation of the parameter *area ratio*. (A) Cell contour detected initially and the area enclosed by this contour. (B) Cell contour encircled by its convex hull and the area enclosed by the convex hull. (C) Graphical representation of the area difference between convex hull and the initial contour. (D and E) Example of the area difference defined in (C) corresponding to an *area ratio* below 1.05 (D) and above 1.05 (E).

- Aspect ratio

Aspect ratio refers to the ratio between the width and the height of the cell contour's bounding box (*x*-size/*y*-size). It can assist in excluding highly elongated objects, such as cell doublets or red blood cells (RBCs), from the analysis. In samples such as whole blood or bone marrow, exclusion of abundant RBCs ($\sim$1000 times the number of WBCs) through *aspect ratio* gating is advisable if RBCs are not of interest for later analysis. Not only *ShapeOut* but also the measurement software *ShapeIn* allows for filtering based on *aspect ratio* of the contour's bounding box. Due to their shape and their softness, RBCs have very large aspect ratios compared to white blood cells. Filtering for aspect ratio values between 0.5 and 2 effectively excludes RBCs from the measurement. If this is done during data acquisition in *ShapeIn*, recording of excessive RBC data, that would be discarded later, is avoided.

4.2 COMPARING MECHANICAL PROPERTIES OF CELL POPULATIONS

The mechanical fingerprint of a cell population is plotted in a deformation vs. cell size scatter plot that has become a common way to present and compare RT-DC-generated data. It is important to note that hydrodynamic shear and normal stresses on the cell depend on cell size (Mietke et al., 2015; Mokbel et al., 2017; Otto et al., 2015). Larger cells leave a thinner film of liquid between the cell membrane and the microfluidic channel wall, such that the shear rate and the resulting shear stresses on the cell are higher than those for small cells at the same overall flow rate. To take this into account, so-called "isoelasticity" lines drawn into the deformation vs. area plots, as shown in Fig. 1C. If a large cell is deformed more as compared to a smaller cell, it does not necessarily mean that it is softer.

By using a model assuming laminar flow with a homogeneous and isotropic elastic sphere as a cell and performing numerical simulations, a lookup table for apparent Young's moduli, as graphically represented in Fig. 5B, was determined specifically for RT-DC experiments (Mokbel et al., 2017). This table contains values corrected for effects such as shear-thinning of the MB and image pixelation discussed elsewhere (Herold, 2017). The program *ShapeOut* includes the lookup table and allows for the transformation of cell deformation and size data into apparent Young's modulus values. Especially in cases where cells of heterogeneous sizes (e.g., from cell culture) are being compared, this additional step can help to reveal mechanical differences. A demonstration of a Young's modulus transformation is depicted in Fig. 5, which shows a comparison of the mechanical properties for a population of neural progenitor cells (NPCs), embryonic stem cells (ESCs), and induced pluripotent stem cells (iPSCs) from mouse (Urbanska et al., 2017). The population of iPSCs has clearly larger deformation values as compared to the population of ESCs, which are slightly smaller in size (Fig. 5A). However, if their Young's modulus is considered, it becomes evident that both cell populations have comparable stiffness (Fig. 5C). Moreover, NPC and iPSC cell populations with corresponding deformation values (Fig. 5A) have, in fact, different Young's moduli (Fig. 5C).

While translating the results into Young's modulus, one has to keep in mind that the lookup table used for determining these values is applicable only for cells that are initially spherical (i.e., have negligible deformation in the "Reservoir" region of the chip) and covers a limited area within the deformation vs. cell size space (Fig. 5B). Evaluating samples that lie on the boundaries of this table may result in discarding significant percentages of data and bias the outcome.

4.3 STATISTICAL ANALYSIS

Due to the large number of events collected in each RT-DC measurement, and the resulting low standard error of the mean, comparing two different cell populations with a standard statistical test such as Student's t-test or Mann–Whitney U test becomes biased toward overestimated significance. With these tests, it is not uncommon to obtain significant differences when comparing two measurements of an identical sample. These differences, however, do not represent actual changes in cell properties but originate from measurement noise caused by biological and technical factors such as day-to-day sample variations, temperature, sample handling, and variations in chip geometry among others.

To overcome this problem, a statistical model considering the variations due to noise separately from the differences caused by treatments or sample type is needed. A well-suited model for statistical evaluation of changes observed with RT-DC, applied previously in a number of studies (Herbig et al., 2017; Kräter et al., 2017; Rosendahl et al., 2018; Toepfner et al., 2018; Urbanska et al., 2017), is a linear mixed

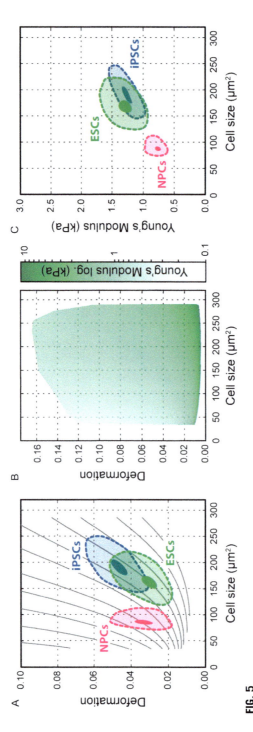

FIG. 5

Determination of Young's modulus from raw RT-DC data. (A) Deformation vs. cell size contour plots of a measurement performed on neural progenitor cells (NPCs, magenta), induced pluripotent stem cells (iPSCs, blue), and embryonic stem cells (ESCs, green) from mouse. Outer contours delineate 50% density (dashed lines), while smaller contours (filled areas) correspond to 95% event density. Isoelasticity lines derived from the numerical simulations are indicated in gray. (B) A graphical representation of the simulation-based lookup table that specifies a Young's modulus value for each point of the colored region in the plot and is used for transformation of deformation and cell size data into Young's modulus. (C) Young's modulus obtained by using lookup table represented in (B) plotted against cell size. Cell types and contour densities as in (A).

model with random intercept and random slope. This model sets the random variation caused by noise into relation to the difference that might be induced by the treatment and estimates a meaningful *P*-value. The linear mixed model analysis implemented in *ShapeOut* under the tab "Analysis" is based on the lme4 package (Bates, Mächler, Bolker, & Walker, 2014) for R (R Core Team 2015) and discussed in detail elsewhere (Herbig et al., 2017). In *ShapeOut*, the loaded measurements can be assigned to the treatment and control groups. Linear mixed model analysis can be applied to any of the recorded parameters.

4.4 MULTIPARAMETRIC CELL PHENOTYPING

Thanks to the stored recordings of cell images, further image-based cell parameters, such as cell brightness or its standard deviation which gives an estimate of cell granularity, are accessible. Along with the standard contour-based parameters, they can facilitate improved discrimination between different cell types. An excellent example is the analysis of nucleated cells in whole blood (Toepfner et al., 2018), where the parameter *brightness* allows for identification of all leukocyte subtypes (lymphocytes, neutrophils, eosinophils, basophils, monocytes). Such separation is impossible to achieve using the parameters *deformation* and *cell size* alone. Raw image data for each cell event are stored in the RT-DC data in standard ".avi" files and are accessible for further custom analysis by the user. The possibilities range from manual inspection to automated algorithmic analysis. We only begin to explore the resolving power of the additional image-based parameters, which, together with machine learning algorithms, will enable efficient identification of cell types in mixed populations.

An additional important parameter of interest for many biological experiments is fluorescence. Discrimination based on fluorescence has become indispensable not only for characterization of gene expression via fluorescent protein-based reporters but also for cell identification by extrinsic labels such as cell permeant stains or antibody conjugates. Recently a tool for simultaneous mechanical characterization and flow cytometry-like fluorescence characterization called real-time fluorescence and deformability cytometry (RT-FDC) was introduced and its use for specific phenotyping of subpopulations defined by fluorescence was demonstrated (Rosendahl et al., 2018; Urbanska et al., 2017). Examples of RT-FDC applications include cell-cycle specific mechanical characterization without the need for prior synchronization or sorting, and mechanical characterization of CD34-positive stem and progenitor cells in a heterogeneous sample.

5 DISCUSSION

RT-DC is a versatile tool for rapid and continuous mechanical characterization of large populations of suspended cells with real-time data evaluation capacity and integrated functionality to extract the cells' Young's modulus (Mietke et al., 2015;

Mokbel et al., 2017; Otto et al., 2015). The well-established experimental procedures summarized in this chapter, together with automated data acquisition and low sample volume required for analysis, render RT-DC an attractive method for both basic research (Kräter et al., 2017; Munder et al., 2016; Rosendahl et al., 2018; Tavares et al., 2017; Urbanska et al., 2017) as well as medical applications (Guzniczak et al., 2017; Koch et al., 2017; Toepfner et al., 2018; Xavier et al., 2016).

One feature unique to our approach is the real-time data evaluation on-the-fly which is of particular value, as it opens the door to the implementation of mechanics-based cell sorting. The cell deformation could be used as a gating parameter to separate cells of desired mechanical properties using one of many microfluidics approaches to sort cells in an active manner (Nawaz et al., 2015; Zhu & Trung Nguyen, 2010). Such separation would enable mechanics-selected cells to be used for downstream analysis or for translational purposes.

Another invaluable asset of RT-DC is its conjunction with an analytical model (Mietke et al., 2015) and numerical simulations (Mokbel et al., 2017) that allows for mapping of cell deformation to the apparent Young's modulus as an inherent material property. This is especially useful for quantitative comparison of mechanical properties of cells with different sizes (see Fig. 5) and for comparison of results obtained with other methods. It is important to note, however, that both theoretical approaches require simplified assumptions that may not fully reflect the structural nature of the cells, such as treating them as homogeneous and isotropic elastic bodies that are spherical before entering the channel. Such assumptions are equivalent to those customarily employed in AFM-enabled nano-indentation when extracting an elastic modulus using the Hertz model or its derivatives. Regardless of these simplifications, the experimentally observed cell deformations correspond well with the results obtained by modeling and simulation. Therefore, the assigned Young's moduli can serve as an effective parameter facilitating the comparison of cell stiffness. In the future, the availability of elastic cell mechanical mimics could prove useful for effective comparison between various cell mechanical measurement techniques, as well as inherent calibration standard prior to RT-DC measurement (Girardo et al., 2018).

Our platform performs mechanical characterization at rates of up to 1000 cells per second while being sensitive for changes in actin organization. The high throughput of microfluidic approaches provides evident advantages such as reduction of measurement time, power to resolve heterogeneities within large cell populations and, in case of mixed samples, the ability to yield sufficient cell numbers to probe rare cell types. It comes, however, with the drawback of deforming cells within short times and, thus, at high rates. Living cells are complex materials that exhibit mechanical properties dependent on strain magnitude and rate. Very high strain rates can potentially lead to fluidization of actin networks and, thus, predominantly probe the viscous response of the cell (Fabry et al., 2001; Hoffman & Crocker, 2009; Kollmannsberger & Fabry, 2011). In RT-DC, cells are deformed within a few milliseconds by forces on the order of sub-μN (Mietke et al., 2015; Otto et al., 2015). This places the method at an intermediate position among other microfluidic approaches such as DC (Gossett et al., 2012) which operates at microsecond time-scales with forces exceeding 1 µN on one end, and quantitative deformability cytometry (qDC) (Nyberg et al., 2017) or suspended microchannel resonator (SMR) (Byun et al., 2013) which operate at the timescales of seconds with pN forces on the other. Contrary to DC which operates at very high strain rates and is sensitive to cytoplasm viscosity and nuclear stiffness (Gossett et al., 2012), RT-DC is primarily sensitive to cytoskeletal changes, in particular in actin organization, as shown by drug screening assays (Golfier et al., 2017). Methods based on pressure-driven passage of cells through a constriction smaller than the cell size such as SMR (Byun et al., 2013), microconstriction arrays (Lange et al., 2015), or qDC (Nyberg et al., 2017) are also sensitive to actin cytoskeleton remodeling. They have, however, lower throughput and bring the cells in contact with channel walls which may introduce the effects of friction to the measurements. Although, when probing of slow responses is of primary interest, such transit methods, together with other classical methods operating at longer timescales, such as OS and AFM, may be recommendable. An appropriate choice of methods for probing mechanical properties at timescale relevant to processes of interest, such as cell circulation and migration, has been exploited in the functional characterization of human myeloid precursor cells progeny (Ekpenyong et al., 2012).

High-throughput experimental approaches to study cell mechanics such as RT-DC have the power to not only accelerate the establishment of cell stiffness as a label-free marker of cell state in physiological processes and diseases, but also to identify molecular contributors to the cell mechanical phenotype. RT-DC methodology was proven successful in discriminating a plethora of different cellular conditions such as yeast dormancy (Munder et al., 2016), stem cell potency and differentiation (Guzniczak et al., 2017; Kräter et al., 2017; Otto et al., 2015; Urbanska et al., 2017; Xavier et al., 2016), malignant transformation (Tavares et al., 2017), activation of the immune system (Toepfner et al., 2018), and malaria infection (Koch et al., 2017), among others. Apart from the characterization of the mechanical properties of cellular states, RT-DC can also be leveraged to dissect molecular candidates underpinning the mechanical differences between cells. Such dissection can be realized by applying drugs or siRNA libraries to cells and screening for phenotype changes. The latter approach, together with fluorescence-based identification of cell-cycle phase, has been demonstrated in an RNAi screen for cell mechanics regulators during mitosis (Rosendahl et al., 2018). Future integration of sample handling automation could make the screening approaches more accessible and facilitate discoveries of novel contributors to cell stiffness. Also, current machine learning approaches in conjunction with various omics datasets (Ciucci et al., 2017) will be increasingly useful in the regard of the dissection of molecular contributors to cell mechanics.

Taken together, the hallmarks of RT-DC such as contactless operation, continuous real-time data processing, high-throughput, sensitivity to cytoskeletal changes, and availability of powerful tools to analyze the manifold contour and image parameters and to infer inherent mechanical properties make it a unique microfluidic platform for mechanical characterization of cells. New functionalities such as simultaneous fluorescence detection (Rosendahl et al., 2018) and future developments such as downstream sorting or automation of sample handling will foster discoveries and contribute to the understanding of the role of cell mechanics in disease, development, and other biological processes.

ACKNOWLEDGMENTS

The authors would like to thank Georg Krainer for fruitful discussions and useful comments on the manuscript. We further thank Zellmechanik Dresden for providing materials for graphics, and the Microstructure Facility at the Center for Molecular and Cellular Bioengineering (CMCB) at Technische Universität Dresden (in part funded by the State of Saxony and the European Regional Development Fund) and Alejandro Riviera Prieto for help with the production of RT-DC chips. This work was financially supported by the Alexander von Humboldt-Stiftung (Alexander von Humboldt Professorship to J.G.).

REFERENCES

Bates, D., Mächler, M., Bolker, B., & Walker, S. (2014). Fitting linear mixed-effects models using lme4. *Journal of Statistical Software*, *67*(1), 1–48. https://doi.org/10.18637/jss.v067.i01.

Byun, S., Son, S., Amodei, D., Cermak, N., Shaw, J., Kang, J. H., et al. (2013). Characterizing deformability and surface friction of cancer cells. *Proceedings of the National Academy of Sciences of the United States of America*, *110*(19), 7580–7585. https://doi.org/10.1073/pnas.1218806110.

Ciucci, S., Ge, Y., Durán, C., Palladini, A., Jiménez-Jiménez, V., Martínez-Sánchez, L. M., et al. (2017). Enlightening discriminative network functional modules behind principal component analysis separation in differential-omic science studies. *Scientific Reports*, *7*, 1–24. https://doi.org/10.1038/srep43946.

Di Carlo, D. (2012). A mechanical biomarker of cell state in medicine. *Journal of Laboratory Automation*, *17*(1), 32–42. https://doi.org/10.1177/2211068211431630.

Ekpenyong, A. E., Whyte, G., Chalut, K., Pagliara, S., Lautenschläger, F., Fiddler, C., et al. (2012). Viscoelastic properties of differentiating blood cells are fate- and function-dependent. *PLoS One:7*(9), e45237. https://doi.org/10.1371/journal.pone.0045237.

Fabry, B., Maksym, G. N., Butler, J. P., Glogauer, M., Navajas, D., & Fredberg, J. J. (2001). Scaling the microrheology of living cells. *Physical Review Letters*, *87*(14), 1–4. https://doi.org/10.1103/PhysRevLett.87.148102.

Girardo, S., Träber, N., Wagner, K., Gheorghe, C., Herold, C., Goswami, R., et al. (2018). Standardized microgel beads as elastic cell mechanical probes. *BioRxiv*. https://doi.org/10.1101/290569.

Golfier, S., Rosendahl, P., Mietke, A., Herbig, M., Guck, J., & Otto, O. (2017). High-throughput cell mechanical phenotyping for label-free titration assays of cytoskeletal modifications. *Cytoskeleton*, *74*(8), 283–296. https://doi.org/10.1002/cm.21369.

Gossett, D. R., Tse, H. T. K., Lee, S. a., Ying, Y., Lindgren, A. G., Yang, O. O., et al. (2012). Hydrodynamic stretching of single cells for large population mechanical phenotyping.

Proceedings of the National Academy of Sciences of the United States of America, *109*(20), 7630–7635. https://doi.org/10.1073/pnas.1200107109.

Guck, J., Ananthakrishnan, R., Mahmood, H., Moon, T. J., Cunningham, C. C., & Käs, J. (2001). The optical stretcher: A novel laser tool to micromanipulate cells. *Biophysical Journal*, *81*(2), 767–784. https://doi.org/10.1016/S0006-3495(01)75740-2.

Guzniczak, E., Mohammad Zadeh, M., Dempsey, F., Jimenez, M., Bock, H., Whyte, G., et al. (2017). High-throughput assessment of mechanical properties of stem cell derived red blood cells, toward cellular downstream processing. *Scientific Reports*, *7*(1), 14457. https://doi.org/10.1038/s41598-017-14958-w.

Herbig, M., Kräter, M., Plak, K., Müller, P., Guck, J., & Otto, O. (2017). *Real-time deformability cytometry: Label-free functional characterization of cells. In R. Hawley & T. Hawley (Eds.), Flow cytometry protocols* (4th ed.). New York, NY, USA: Humana Press. https://doi.org/10.1007/978-1-4939-7346-0.

Herold, C. (2017). Mapping of deformation to apparent Young's modulus in real-time deformability. *Cytometry*. Retrieved from http://arxiv.org/abs/1704.00572.

Hochmuth, R. M. (2000). Micropipette aspiration of living cells. *Journal of Biomechanics*, *33*(1), 15–22. https://doi.org/10.1016/S0021-9290(99)00175-X.

Hoffman, B. D., & Crocker, J. C. (2009). Cell mechanics: Dissecting the physical responses of cells to force. *Annual Review of Biomedical Engineering*, *11*, 259–288. https://doi.org/10.1146/annurev.bioeng.10.061807.160511.

Koch, M., Wright, K. E., Otto, O., Herbig, M., Salinas, N. D., Tolia, N. H., et al. (2017). *Plasmodium falciparum* erythrocyte-binding antigen 175 triggers a biophysical change in the red blood cell that facilitates invasion. *Proceedings of the National Academy of Sciences of the United States of America*, *114*(16), 4225–4230. https://doi.org/10.1073/pnas.1620843114.

Kollmannsberger, P., & Fabry, B. (2011). Linear and nonlinear rheology of living cells. *Annual Review of Materials Research*, *41*(1), 75–97. https://doi.org/10.1146/annurev-matsci-062910-100351.

Kräter, M., Jacobi, A., Otto, O., Tietze, S., Müller, K., Poitz, D. M., et al. (2017). Bone marrow niche-mimetics modulate HSPC function via integrin signaling. *Scientific Reports*, *7*(1), 2549. https://doi.org/10.1038/s41598-017-02352-5.

Lange, J. R., Steinwachs, J., Kolb, T., Lautscham, L. A., Harder, I., Whyte, G., et al. (2015). Microconstriction arrays for high-throughput quantitative measurements of cell mechanical properties. *Biophysical Journal*, *109*(1), 26–34. https://doi.org/10.1016/j.bpj.2015.05.029.

Maloney, J. M., Nikova, D., Lautenschläger, F., Clarke, E., Langer, R., Guck, J., et al. (2010). Mesenchymal stem cell mechanics from the attached to the suspended state. *Biophysical Journal*, *99*(8), 2479–2487. https://doi.org/10.1016/j.bpj.2010.08.052.

Mietke, A., Otto, O., Girardo, S., Rosendahl, P., Taubenberger, A., Golfier, S., et al. (2015). Extracting cell stiffness from real-time deformability cytometry: Theory and experiment. *Biophysical Journal*, *109*(10), 2023–2036. https://doi.org/10.1016/j.bpj.2015.09.006.

Mokbel, M., Mokbel, D., Mietke, A., Träber, N., Salvatore, G., Otto, O., et al. (2017). Numerical simulation of real-time deformability cytometry to extract cell mechanical properties. *ACS Biomaterials Science & Engineering*, *3*(11), 2962–2973. https://doi.org/10.1021/acsbiomaterials.6b00558.

Munder, M. C., Midtvedt, D., Franzmann, T., Nüske, E., Otto, O., Herbig, M., et al. (2016). A pH-driven transition of the cytoplasm from a fluid- to a solid-like state promotes entry into dormancy. *eLife*, *5*, e09347. https://doi.org/10.7554/eLife.09347.

Nawaz, A. A., Chen, Y., Nama, N., Nissly, R. H., Ren, L., Ozcelik, A., et al. (2015). Acousto-fluidic fluorescence activated cell sorter. *Analytical Chemistry*, *87*(24), 12051–12058. https://doi.org/10.1021/acs.analchem.5b02398.

Nematbakhsh, Y., & Lim, C. T. (2015). Cell biomechanics and its applications in human disease diagnosis. *Acta Mechanica Sinica*, *31*(2), 268–273. https://doi.org/10.1007/s10409-015-0412-y.

Nyberg, K. D., Hu, K. H., Kleinman, S. H., Khismatullin, D. B., Butte, M. J., & Rowat, A. C. (2017). Quantitative deformability cytometry: Rapid, calibrated measurements of cell mechanical properties. *Biophysical Journal*, *113*(7), 1574–1584. https://doi.org/10.1016/j.bpj.2017.06.073.

Otto, O., Rosendahl, P., Mietke, A., Golfier, S., Herold, C., Klaue, D., et al. (2015). Real-time deformability cytometry: On-the-fly cell mechanical phenotyping. *Nature Methods*, *12*(3), 199–202. https://doi.org/10.1038/nmeth.3281.

Radmacher, M. (2007). Studying the mechanics of cellular processes by atomic force microscopy. *Methods in Cell Biology*, *83*(7), 347–372. https://doi.org/10.1016/S0091-679X(07)83015-9.

Rosendahl, P., Plak, K., Jacobi, A., Kräter, M., Töpfner, N., Otto, O., et al. (2018). Real-time fluorescence and deformability cytometry. *Nature Methods*, *15*(5), 355–358. https://doi.org/10.1038/nmeth.4639.

Silverthorn, D. U., & Johnson, B. R. (2009). *Human physiology an integrated approach*. San Francisco, CA, USA: Pearson.

Tavares, S., Vieira, A. F., Taubenberger, A. V., Araújo, M., Martins, N. P., Brás-Pereira, C., et al. (2017). Actin stress fiber organization promotes cell stiffening and proliferation of pre-invasive breast cancer cells. *Nature Communications: 8*. https://doi.org/10.1038/ncomms15237 (online publication).

Toepfner, N., Herold, C., Otto, O., Rosendahl, P., Jacobi, A., Kräter, M., et al. (2018). Detection of human disease conditions by single-cell morpho-rheological phenotyping of blood. *eLife, 7*, e29213. https://doi.org/10.7554/eLife.29213.

Urbanska, M., Winzi, M., Neumann, K., Abuhattum, S., Rosendahl, P., Müller, P., et al. (2017). Single-cell mechanical phenotype is an intrinsic marker of reprogramming and differentiation along the mouse neural lineage. *Development*, *144*(23), 4313–4321. https://doi.org/10.1242/dev.155218.

Xavier, M., Rosendahl, P., Herbig, M., Kräter, M., Spencer, D., Bornhäuser, M., et al. (2016). Mechanical phenotyping of primary human skeletal stem cells in heterogeneous populations by real-time deformability cytometry. *Integrative Biology*, *8*(5), 616–623. https://doi.org/10.1039/c5ib00304k.

Zhu, G., & Trung Nguyen, N. (2010). Particle sorting in microfluidic systems. *Micro and Nanosystems*, *2*(3), 202–216. https://doi.org/10.2174/1876402911002030202.

A microfluidic cell-trapping device to study dynamic host-microbe interactions at the single-cell level

11

Chiara Toniolo[1], Matthieu Delincé[2], John D. McKinney

School of Life Sciences, Swiss Federal Institute of Technology in Lausanne (EPFL),
Lausanne, Switzerland
[1]*Corresponding author: e-mail address: chiara.toniolo@epfl.ch*

CHAPTER OUTLINE

[2]Current address: HiFiBiO Inc, Cambridge, MA, United States

Methods in Cell Biology, Volume 147, ISSN 0091-679X, https://doi.org/10.1016/bs.mcb.2018.06.008

Abstract

Single-cell imaging of host-microbe interactions over time is impeded by cellular motility because the cells under scrutiny tend to migrate out of the imaging field. To overcome this technical challenge, we developed a microfluidic platform for imaging hundreds of individual motile phagocytic cells and bacteria within microfluidic traps that restrict their movement. The interaction of trapped host cells and bacteria is monitored by long-term time-lapse microscopy, allowing direct visualization of all stages of infection at the single-cell level. The medium flowing through the microfluidic device can be changed quickly and precisely, permitting the real-time imaging of cellular responses to antibiotics or other environmental stresses. Here, we demonstrate the potential applications of this approach by co-culturing the phagocytic amoeba *Dictyostelium discoideum* with the intracellular pathogen *Mycobacterium marinum*. However, the platform can be adapted easily for use with other host cells or microorganisms. This approach will provide new insights into host-pathogen interactions that cannot be studied using conventional population-based assays.

1 INTRODUCTION

The encounter of a phagocytic cell and a bacterium usually culminates in phagocytic uptake and destruction of the invading microorganism. However, some pathogenic bacteria have acquired mechanisms to survive and sometimes replicate intracellularly, which may result in death of the host cell (Finlay & McFadden, 2006; Jayachandran, BoseDasgupta, & Pieters, 2013). In recent years, it has been established that there exist marked cell-to-cell differences within clonal cell populations. This intrinsic "phenotypic heterogeneity" may allow subpopulations of bacteria to survive host-imposed stresses and establish an infection (Ackermann, 2015; Helaine et al., 2014; Wakamoto et al., 2013). The majority of studies on host-pathogen interactions are carried out using animal models or *in vitro* batch cultures. Typical readouts of these approaches are host survival and quantification of bacterial burdens by counting colony-forming units. Unfortunately, these assays provide only population-averaged data and little information about cell-to-cell heterogeneity and temporal changes at the host and pathogen level. Single-cell techniques such as immunofluorescence microscopy or flow cytometry can provide "snapshots" of cell-to-cell phenotypic heterogeneity. However, these assays do not provide information on the dynamics of host-pathogen interactions because individual bacteria and host cells cannot be tracked over time (Avraham et al., 2015; Bhaskar et al., 2014; Helaine et al., 2014).

Live-cell time-lapse microscopy is a powerful approach to characterize phenotypic heterogeneity in cellular populations and to investigate the impact of this heterogeneity on survival under stressful conditions (Locke & Elowitz, 2009). However, this approach is not used routinely to investigate host-pathogen interactions due to the inherent technical challenges associated with this technique. A major impediment to temporal tracking of both host cells and bacteria is the high motility of phagocytic host cells, which tend to migrate out of the field of view during the course of an experiment. Conventionally, this problem has been addressed by physically tracking individual host cells over time. This approach is labor-intensive because the experimenter must be present throughout the experiment in order to keep the target cell within the imaging field by adjusting the microscope stage. Also, the throughput of this technique is usually limited to just one motile cell per experiment.

As an alternative to physical *tracking* of host-pathogen interactions, in this chapter we describe a novel microfluidic platform suitable for long-term *trapping* and live-cell microscopy of bacteria co-cultured with motile eukaryotic phagocytes. Our approach allows the experimenter to image hundreds of host cells in parallel over time by trapping them in micro-chambers where they are free to move, grow, and divide with normal kinetics, yet are prevented from exiting the field of view. The microfluidic device is designed to allow continuous flow of culture medium and thus permits rapid and reversible medium switches ideal for real-time analysis of cellular responses and adaptations to fluctuating environments, for example, addition or removal of antibiotics or other stressors.

Here, we use the new microfluidic cell-trapping device to demonstrate long-term time-lapse microscopy of the motile amoeba *Dictyostelium discoideum*, an important model for experimental analysis of host-microbe interactions (Cosson & Soldati, 2008), in co-culture with *Mycobacterium marinum*, an intracellular fish pathogen that is closely related to the human pathogen *Mycobacterium tuberculosis* (Stinear et al., 2008). However, our cell-trapping platform and methods could be readily adapted to study different organisms, for example, interactions between mammalian macrophages and *M. tuberculosis*.

2 PROTOCOL

The microfluidic platform described here (Delincé et al., 2016) is based on previous designs (Wakamoto et al., 2013; Wakamoto, Inoue, Moriguchi, & Yasuda, 2001) and comprises three parts: a polydimethylsiloxane (PDMS) microfluidic chip, a semi-permeable membrane, and a patterned glass coverslip (Fig. 1A). The three parts are clamped together by a polymethyl methacrylate (PMMA)-aluminum holder (Fig. 1C). We use photolithography to pattern one face of the coverslip with thousands of circular culturing chambers fabricated from SU8, in which cells can be trapped and imaged over time (Fig. 1A–B). The microfluidic chip on the top of the device is connected to a syringe controlled by a syringe pump, which allows

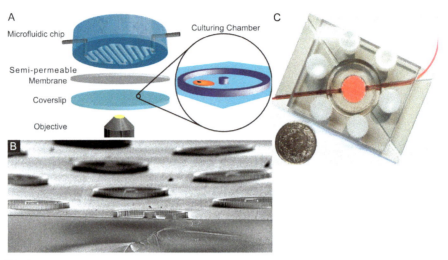

FIG. 1

The microfluidic platform. (A) Blow-up schematic of the microfluidic device, which consists of a PDMS chip, a semi-permeable membrane, and a micro-patterned coverslip.
(B) Scanning electron microscope image of a patterned coverslip showing the SU8 walls forming the circular traps micro-chambers. (C) Top view photograph of the microfluidic device clamped in its PMMA-aluminum holder. A red food dye was injected in the chip to better visualize the microfluidic channels. A one Swiss Franc coin is pictured for scale.

Reproduced and adapted from Delincé, M. J., Bureau, J.-B., López-Jiménez, A. T., Cosson, P., Soldati, T., & McKinney, J. D. (2016). A microfluidic cell-trapping device for single-cell tracking of host–microbe interactions. Lab on a Chip, 17, 32–46. https://doi.org/10.1039/C6LC00649C with permission from the Royal Society of Chemistry.

for a continuous flow of medium that can be rapidly switched if needed. The semi-permeable membrane sandwiched between the flow channels and the micro-chambers on the glass coverslip allows the medium to diffuse into the chambers but prevents the cells from being washed out of the chambers by the flow of medium. This setup also prevents motile cells from migrating out of the circular micro-chambers (traps).

2.1 DESIGN OF MICROFLUIDIC PLATFORM

2.1.1 Micro-patterned coverslip design

The culturing micro-chamber architecture can be designed using a layout editing software (e.g., L-edit, Tanner EDA) or a CAD software (e.g., AutoCAD Autodesk). The design can then be either transferred onto a chromium mask by laser writing or printed on a plastic photomask. The masks can be made in a microfabrication facility or purchased commercially (e.g., http://outputcity.com). When designing the culturing chamber, some principles should be taken into consideration:

- The size of the chamber should be optimized based on the size of the field of view of the camera and the magnification of the objective used.
- The chambers should be wide and tall enough to allow the cells to move freely within the traps without being squeezed.
- To prevent the membrane from collapsing into the chambers and squeezing the cells, an appropriate number of pillars to support the membrane should be added to the chambers. The optimum number of pillars would depend on the area and the height of the chamber.

Here, we used the motile amoeba *D. discoideum*, which is approximately 10 μm in diameter and 5 μm in height (Delincé et al., 2016). We designed circular culturing micro-chambers measuring 70 μm in diameter and 4.1 μm in height (Fig. 2A).

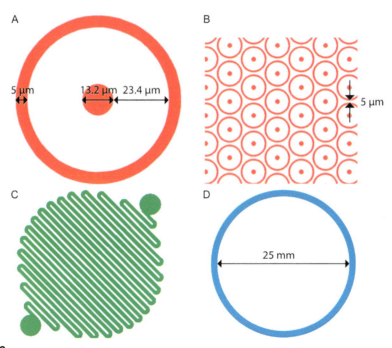

FIG. 2

Microfluidic platform and mold designs. (A) Top view of a single circular micro-chamber with its dimensions. (B) Geometrical organization of the chambers on the coverslip. Chambers are equidistant and 5 μm apart. (C) Microfluidic channel design consisting of a single-layer PDMS chip with a serpentine flow channel 300 μm wide and 300 μm high. The inlet and outlet tubing are connected to the two circular chambers at the extremities of the flow channel. (D) Top view and dimensions of the circular mold used to prepare the semi-permeable agarose membrane.

Reproduced and adapted from Delincé, M. J., Bureau, J.-B., López-Jiménez, A. T., Cosson, P., Soldati, T., & McKinney, J. D. (2016). A microfluidic cell-trapping device for single-cell tracking of host–microbe interactions. Lab on a Chip, 17, 32–46. https://doi.org/10.1039/C6LC00649C with permission from the Royal Society of Chemistry.

One pillar measuring 13.2 μm in diameter was added in the center of each chamber to prevent the membrane from collapsing. The circular chambers are positioned 5 μm apart (Fig. 2B). The close proximity between chambers reduces the total distance that the microscope stage must travel when performing multi-point time-lapse imaging. This increases stability, ensures the maintenance of optical focus over time, and allows faster cycles of multi-chamber imaging. The dimensions of the device described here were optimized for imaging *D. discoideum* with a 100 × oil-immersion objective. However, these parameters could be modified readily to accommodate different experimental variables, such as cell size, microscope objective, and camera image size.

2.1.2 Microfluidic chip design and fabrication of molds

Molds can be fabricated by various microfabrication techniques (e.g., photolithography) and can be ordered from commercial suppliers (e.g., http://ufluidix.com). Two molds are required to prepare the microfluidic platform: the first mold is used to fabricate the microfluidic chip; the second mold is needed to prepare the semi-permeable membrane.

The design for the microfluidic chip described here consists of a single serpentine channel 300 μm wide and 300 μm high (Fig. 2C). This single-channel architecture reduces the formation of air bubbles that can appear at channel bifurcations.

The semi-permeable membrane used in the microfluidic platform should match the diameter of the glass coverslip. We use a mold consisting of a circular architecture 25 mm large and 200 μm high to prepare it (Fig. 2D).

2.2 FABRICATION OF MICRO-PATTERNED COVERSLIPS

2.2.1 Materials

- 25 mm coverslips, Menzel-Gläser, 25 mm diameter, #1
- SU8 1060 Gersteltech
- Mask (see Section 2.1.1)
- Propylene glycol monomethyl ether acetate (PGMEA)
- Isopropanol

2.2.2 Equipment

- Plasma
- Spin coater
- Hot plate
- Mask aligner
- Profilometer
- Clean compressed nitrogen or air source

2.2.3 Method

1. Clean coverslip by oxygen plasma treatment for 7 min at 500 W with 400 mL/min O_2.
2. Coat coverslip with resin. Center the coverslip on the spin coater chuck, place a drop of SU8 1060 resin on the coverslip, and start spinning coating. Spin-coating recipe: 5 s to 500 rpm; 5 s at 500 rpm; 15 s to 5900 rpm; 30 s at 5900 rpm; 20 s to 0 rpm.
3. Soft-bake the SU8 on a hot plate. Bring to 120 °C in 10 min, keep at 120 °C for 5 min; bring back to room temperature in 1 h.
4. Expose photoresist using a UV mask aligner with an exposure dose of 150 mJ/cm^2 and the chamber design mask.
5. Post-exposure bake: bring to 65 °C in 10 min, stay at 65 °C for 10 min, bring to 95 °C in 15 min, stay at 95 °C for 15 min, cool down to room temperature in 1 h.
6. Develop photoresist by dipping the coverslip in PGMEA for 90 s. Rinse in isopropanol for 30 s and dry coverslips by blowing nitrogen.
7. Hard bake the resin on a hot plate: bring to 105 °C in 15 min, stay at 105 °C for 15 min, cool down to room temperature in 1 h.
8. Confirm thickness with a profilometer. The procedure should yield a thickness of 4.1 μm.

2.3 FABRICATION OF AGAROSE SEMI-PERMEABLE MEMBRANE

2.3.1 Materials

- Agarose
- PBS or culturing medium
- Mold (see Section 2.1.2)
- Glass wafer
- Plastic-tipped tweezers

2.3.2 Method

1. Prepare a 2% agarose mix of culturing medium without fetal bovine serum.
2. Microwave until agarose is completely dissolved.
3. Pour the mixture onto a mold to make 25 mm diameter and 200 μm high agarose pads.
4. Place a glass wafer on top.
5. Leave to solidify for 10 min.
6. Detach the agarose pads from the mold using plastic-tipped tweezers. Keep them moist by placing them in a small Petri dish with 2 mL of medium.

2.4 FABRICATION OF PDMS MICROFLUIDIC CHIP

2.4.1 Materials

- PDMS elastomer
- Curing agent

- Mold (see Section 2.1.2)
- Silicon tubing

2.4.2 Equipment
- PDMS mixer
- Desiccator
- Oven
- 25 mm puncher
- 2 mm biopsy puncher

2.4.3 Method
1. Mix PDMS elastomer with curing agent in a 10:1 ratio to make a total of 50 g. It can be mixed manually or, for more consistent results, using a planetary mixer (Thinky Mixer ARE-250).
2. Pour the PDMS mixture onto the mold and place under vacuum for 30 min. For faster degassing, first place a thin layer of PDMS on the wafer and degas for 10 min, then add the remaining PDMS and degas for another 20 min.
3. Cure PDMS in an oven at 80 °C for 50 min.
4. Cut the chip using a 25 mm punch.
5. Using a biopsy puncher of 2 mm diameter, punch the two connecting holes.
6. Connect silicon tubing to inlet and outlet of chip (0.76/1.56 mm inner/outer diameter).
7. Seal tubing using uncured PDMS.

2.5 SEEDING CELLS IN THE TRAPPING DEVICE

As an example, we describe here the procedure for seeding the amoeba *D. discoideum* and the bacterium *M. marinum* in the device. However, the design of the microfluidic device can be optimized for use with other organisms (see Section 2.1.1).

2.5.1 Materials
- 5 µm PVDF filter (Millipore)
- 1 mL syringe
- PDMS chip: see Section 2.3
- Patterned coverslip: see Section 2.2
- Agarose semi-permeable membrane: see Section 2.2.1
- Plastic-tipped tweezers
- Holder consisting of a PMMA piece (5 mm thick) and an aluminum one with a 14-mm hole. The two pieces are held together using screws (see Fig. 1C; Dhar & Manina, 2015). The upper PMMA part is transparent to allow transmission microscopy, and the lower aluminum part has an aperture to accommodate the oil-immersion objective.

- *D. discoideum*
- *M. marinum* culture (fluorescent)
- HL5c medium

2.5.2 Equipment

- Sonicator
- Centrifuge
- Desiccator

2.5.3 Method

1. Sonicate PDMS pad for 1 h in isopropanol.
2. Clean tubing and device by flowing 10 mL of 70% ethanol and then rinse by flowing 20 mL of doubled-distilled water. Dry and autoclave for 20 min. Make sure the chip cools down to room temperature before using it. Do not autoclave the patterned coverslip.
3. Prepare the coverslip by putting it in a small petri dish, cover it with 1 mL of medium, and put the dish under vacuum for 5 min. This step will remove air bubbles that might otherwise be trapped in the chambers.
4. Detach *D. discoideum* cells from an 80% confluent culture in a tissue culture dish. Centrifuge 10 mL in a Falcon tube (at 249g for 5 min at room temperature) and resuspend in approximately 100 µL of supernatant. Aspirate medium over the coverslip and add the cells on the coverslip. Let cells settle for at least 10 min.
5. Start with a culture of *M. marinum* in exponential phase (OD$_{600nm}$ between 0.8 and 1.0). Collect the bacteria by centrifuging 6 mL of culture at 1200g for 5 min at room temperature. Wash twice with 6 mL HL5c, repeat centrifugation, and resuspend in 300 µL of the same medium. Filter with a 5 µm PVDF filter into a 1 mL syringe. Remove the plunger and load the syringe with a pipette from the top. This procedure removes bacterial aggregates from the resuspended culture.
6. While the bacteria are being centrifuged, place the membrane on the top of the inverted PDMS microfluidic chip and wait 2 min to let the excess of medium dry out.
7. Spread 5 µL of bacteria on the membrane.
8. Invert the PDMS chip and membrane and center them on the coverslip, then sandwich them between the plexi cover and the machined aluminum plate. Tighten the screws gently, while maintaining some pressure on the central portion of the chip. This procedure prevents the *D. discoideum* cells from flowing out of the chambers.
9. Connect a syringe filled with medium to the tubing of the microfluidic chip.
10. Flow medium through the chip using a syringe pump to ensure that there are no leaks or bubbles forming in the channel.

2.6 IMAGING

Imaging is performed using an inverted epifluorescence microscope equipped with a motorized stage and adapted for live-cell imaging. A motorized stage is required to image multiple positions in the microfluidic device in parallel. For long experiments, it may be important to control the environmental conditions such as temperature, humidity, and CO_2. Objectives with different magnifications and numerical apertures (NAs) can be used, depending on the sample analyzed.

2.6.1 Equipment
- Syringe pump
- Inverted microscope

2.6.2 Method
1. Place the device on the microscope stage, then place the syringe containing the medium in the syringe pump and set the flow rate at $10\,\mu L/min$.
2. Manually select the positions of chambers containing at least one host cell and one bacterium. As a control, select also some chambers containing only bacteria or host cells.
3. Start the automated time-lapse microscopy using appropriate imaging conditions.
4. When the experiment is finished, flow 70% ethanol through the chip, disassemble the device, and autoclave the PDMS chip and tubes. Clean the holder with an appropriate disinfectant. Discard the single-use micro-patterned coverslip and the membrane.

For our setup, we use a DeltaVision Elite inverted microscope equipped with a $100 \times$ NA $= 1.4$ oil objective, a CoolSnap HQ2 camera, and fluorescence filters for GFP (excitation peak 475 nm with 28 nm bandwidth; emission peak 523 nm with 36 nm bandwidth) and for mCherry (excitation peak 542 nm with 27 nm bandwidth, emission peak 594 nm with 45 nm bandwidth). The microscope is enclosed within a custom-fitted plastic box to control the ambient temperature, which is kept constant at $24\,°C$ during the experiments. D. discoideum and M. marinum are imaged on the phase-contrast and fluorescence channels, respectively (Fig. 3A–C). Images are acquired every 1.5 min for 96 h. Typically, 20–50 chambers are imaged per experiment.

2.7 OPTIMIZATION OF TIME-LAPSE IMAGING

The throughput achievable (number of positions imaged in parallel in one experiment) depends on the speed of the camera, the time needed for the automated stage to travel to all of the programmed fields of view, and the interval between two consecutive images required to achieve a sufficient time resolution to investigate the cellular process of interest.

To avoid phototoxicity, it is important to identify a compromise between frequency of imaging (with resultant photodamage) and the time resolution desired.

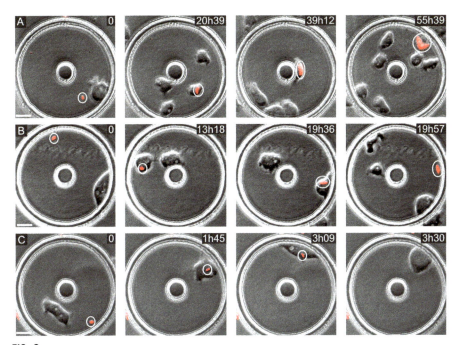

FIG. 3

Time-lapse imaging of co-cultured *D. discoideum* and *M. marinum*. (A) Image series of a co-culture starting with one *D. discoideum* and one *M. marinum* cell constitutively expressing mCherry (in red and highlighted with a white circle). At 20h 39min the *M. marinum* bacterium is internalized by a *D. discoideum* cell. The intracellular bacteria replicate and lyse their host cell at 55h 39min. During this period, *D. discoideum* division events are observed. (B) Image series showing a *D. discoideum* cell that is infected at 13h 18min and then releases the intracellular bacterium at 19h 57min while remaining intact and viable. Some *D. discoideum* division events are observed. (C) Image series showing an *M. marinum* cell that is internalized at 1h 45min and subsequently killed by *D. discoideum* at 3h 30min. (A–C) Scale bar is 10µm.

Reproduced from Delincé, M. J., Bureau, J.-B., López-Jiménez, A. T., Cosson, P., Soldati, T., & McKinney, J. D. (2016). A microfluidic cell-trapping device for single-cell tracking of host–microbe interactions. Lab on a Chip, *17, 32–46. https://doi.org/10.1039/C6LC00649C with permission from the Royal Society of Chemistry.*

This parameter will be dependent on the organism studied, the optical characteristics of the imaging source, the filters on the microscope, and the sensitivity of the camera. Control experiments should be performed to ensure that the host cells and bacteria do not exhibit phototoxicity-induced behaviors different to that observed in conventional batch cultures (e.g., growth rate). If necessary, phototoxicity can be reduced by recording bright-field or phase-contrast images frequently, in order to track the cells' motion over time, while recording fluorescence images less often.

Confocal microscopes offer the ability to do fine optical sectioning and obtain the three-dimensional coordinates of the processes of interest. However, confocal imaging requires multiple exposures at each time point (one for each Z-slice), and therefore increases the risk of phototoxicity. Therefore, in some instances, it may be advisable to use a wide-field microscope, which captures the fluorescence from a larger three-dimensional volume, thus requiring fewer Z stacks and resulting in less phototoxicity.

2.8 EXAMPLES OF IMAGE ANALYSIS

Microscopic image analysis can be carried out using the ImageJ (National Institutes of Health, USA) image analysis software or other available softwares, and should be implemented depending on the specific organism and application. A large number of questions in single-cell infection biology can be explored with the platform and methods described here, and a comprehensive description of cellular properties and processes that could be analyzed is beyond the scope of this chapter. Here, we provide some examples of analysis that we routinely use in our experiments involving *D. discoideum* and *M. marinum*.

2.8.1 Viability of host cells and their division kinetics

The host cells' viability is monitored by visual analysis of the phase-contrast time-lapse series. Death of *D. discoideum* cells is usually preceded by cessation of movement and rounding up of the cell, followed by cellular lysis. The time interval between initial infection and cell death can be calculated as the difference between the first frame where *D. discoideum* is infected and the frame where the host cell dies. For instance, in Fig. 3A one *D. discoideum* cell is infected at 20 h 29 min and dies at 55 h 39 min after initiating the experiment. Similarly, the interdivision time is defined as the time interval from the birth of a cell until its subsequent division (Fig. 4A). Examples of cell divisions are shown in Fig. 3A–B, where the number of *D. discoideum* trapped inside the chambers increases over time. Importantly, the design of the trapping chambers not only prevents the trapped cells from leaving the chamber, it also prevents cells outside the chambers from entering.

2.8.2 Localization, growth, and fate of the bacteria

The localization (intracellular, extracellular, sub-cellular) of the bacteria over time can be analyzed by visual inspection of the fluorescence time-lapse series. Bacteria are considered intracellular when their fluorescence signal overlaps with a *D. discoideum* for several sequential frames. As an example, in Fig. 3A–C, *M. marinum* is extracellular at the first time point shown and is internalized by *D. discoideum* from the second time point onwards. Typically, intracellular *M. marinum* can be killed by the host, released back into the environment, or replicate within the host cell. Bacterial fluorescence is used as a proxy for viability and can be used to monitor bacterial fate. Time-to-death of internalized bacteria is

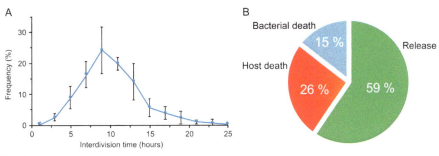

FIG. 4

Examples of cell-to-cell phenotypic heterogeneity in host-pathogen interactions.
(A) Frequency distribution of the interdivision times of *D. discoideum* in the mirofluidic cell-trapping device (mean = 10.2, *n* = 326). Error bars represent the standard deviation of four different experiments analyzed and plotted. (B) Pie chart showing the distribution of the possible outcomes observed in the interaction between *D. discoideum* and *M. marinum*. Following internalization by *D. discoideum*, *M. marinum* can be released without loss of viability to either the host cell or the bacterium (*n* = 136), can be killed by the host cell (*n* = 33), or can replicate intracellularly resulting in death of the host cell (*n* = 66).

Reproduced and adapted from Delincé, M. J., Bureau, J.-B., López-Jiménez, A. T., Cosson, P., Soldati, T., & McKinney, J. D. (2016). A microfluidic cell-trapping device for single-cell tracking of host–microbe interactions. Lab on a Chip, 17, 32–46. https://doi.org/10.1039/C6LC00649C with permission from the Royal Society of Chemistry.

defined as the time interval between the uptake of the bacterium by a host cell and the time point when the bacterium is no longer fluorescent. Fig. 3C shows a representative example of an internalized bacterium killed by *D. discoideum*. Similarly, time-to-release is defined as the time interval between the uptake of the bacterium by a host cell and its release (Fig. 3C). Bacterial growth over time is proportional to the increase in total fluorescence intensity measured in a chamber or overlapping with a *D. discoideum* cell.

3 DISCUSSION

This chapter presented a novel microfluidic cell-trapping platform and methods for trapping, cultivating, infecting, and imaging individual motile eukaryotic cells. Using previous methods, physical tracking of multiple motile cells for extended periods of time was extremely challenging and limited the possibilities for single-cell analysis of host-pathogen interactions. Here, we tested the microfluidic cell-trapping platform using the amoeba *D. discoideum* and we showed that these cells seem to be little affected by trapping in the microfluidic chamber, inasmuch as they display normal viability and growth kinetics. Measurements of cellular interdivision times at the single-cell level confirmed that the average doubling time (10.2 h) (Fig. 4A) is similar to the doubling time of *D. discoideum* cells grown in conventional batch cultures (Fey, Kowal, Gaudet, Pilcher, & Chisholm, 2007). Interestingly, the single-cell interdivision times showed a wide distribution: some cells divide within a couple of hours, while others take more than a day to divide. These observations highlight the existence of significant phenotypic heterogeneity within the population of cells, which could not be measured using conventional, population-based assays.

The tools and methods described here will be particularly useful to study motile phagocytic cells and their interaction with bacteria. Here, we tested it co-culturing *D. discoideum* and the bacterium *M. marinum*. We observed that infection of the motile amoeba *D. discoideum* by the pathogenic bacterium *M. marinum* results in highly variable dynamics and outcomes when analyzed at the single-cell level. This cell-to-cell variation included the unexpected observation that, in a majority of amoeba-bacterium interactions, the internalized bacteria are subsequently released without significantly harming the host. Alternative fates that we observed included intracellular replication of *M. marinum* resulting in death of the host cell or, conversely, killing of the internalized bacterium by the host cell (Fig. 4B). These results demonstrate that changes in total bacterial numbers over time do not reflect intracellular growth alone, but rather a balance of growth, death, and release.

This microfluidic platform can be easily modified and optimized to investigate different cell types and interactions. For instance, single-cell analysis of the interaction of host cells with bacterial mutant strains possessing different levels of virulence could cast new light on the causes and outcomes of the phenotypes observed. Moreover, genetically-encoded fluorescent reporter strains could be used as markers for specific bacterial processes (gene expression, metabolic activity, cell cycle stage...) or host cell functions (acidic pH, reactive oxygen species, reactive nitrogen species...). Within the device, the culturing conditions are precisely controlled by the infused medium, which can be rapidly and precisely switched as needed. Medium switching can also be fully automated using computer-controlled pumps or in-line valve-based switching. It is therefore possible to test multiple culture conditions in a single experiment and observe the effect of drugs or other stresses on the organisms studied in real time.

The tools and methods described here make it possible to generate quantitative single-cell data, to observe how cell-to-cell phenotypic heterogeneity varies dynamically over time, and to monitor the complete sequence of events following initial infection. We are therefore able to map the effects of phenotypic heterogeneity on subsequent outcomes. The ability to generate large-scale single-cell data is essential for systems-level analysis and computational modeling of host-pathogen interactions, which will provide new insights into processes that were previously difficult to investigate.

REFERENCES

Ackermann, M. (2015). A functional perspective on phenotypic heterogeneity in micro-organisms. *Nature Reviews Microbiology*, *13*(8), 497–508. https://doi.org/10.1038/nrmicro3491.

Avraham, R., Haseley, N., Brown, D., Penaranda, C., Jijon, H. B., Trombetta, J. J., et al. (2015). Pathogen cell-to-cell variability drives heterogeneity in host immune responses. *Cell*, *162*(6), 1309–1321. https://doi.org/10.1016/j.cell.2015.08.027.

Bhaskar, A., Chawla, M., Mehta, M., Parikh, P., Chandra, P., Bhave, D., et al. (2014). Reengineering redox sensitive GFP to measure mycothiol redox potential of Mycobacterium tuberculosis during infection. *PLoS Pathogens, 10*(1), e1003902. https://doi.org/10.1371/journal.ppat.1003902.

Cosson, P., & Soldati, T. (2008). Eat, kill or die: When amoeba meets bacteria. *Current Opinion in Microbiology*, *11*(3), 271–276. https://doi.org/10.1016/j.mib.2008.05.005.

Delincé, M. J., Bureau, J.-B., López-Jiménez, A. T., Cosson, P., Soldati, T., & McKinney, J. D. (2016). A microfluidic cell-trapping device for single-cell tracking of host–microbe interactions. *Lab on a Chip*, (17), 32–46. https://doi.org/10.1039/C6LC00649C.

Dhar, N., & Manina, G. (2015). Single-cell analysis of mycobacteria using microfluidics and time-lapse microscopy. *Methods in Molecular Biology (Clifton, N.J.)*, *1285*, 241–256. https://doi.org/10.1007/978-1-4939-2450-9_14.

Fey, P., Kowal, A. S., Gaudet, P., Pilcher, K. E., & Chisholm, R. L. (2007). Protocols for growth and development of Dictyostelium discoideum. *Nature Protocols*, *2*(6), 1307–1316. https://doi.org/10.1038/nprot.2007.178.

Finlay, B. B., & McFadden, G. (2006). Anti-immunology: Evasion of the host immune system by bacterial and viral pathogens. *Cell*, *124*(4), 767–782. https://doi.org/10.1016/j.cell.2006.01.034.

Helaine, S., Cheverton, A. M., Watson, K. G., Faure, L. M., Matthews, S. A., & Holden, D. W. (2014). Internalization of Salmonella by macrophages induces formation of nonreplicating persisters. *Science (New York, N.Y.)*, *343*(6167), 204–208. https://doi.org/10.1126/science.1244705.

Jayachandran, R., BoseDasgupta, S., & Pieters, J. (2013). Surviving the macrophage: Tools and tricks employed by Mycobacterium tuberculosis. *Current Topics in Microbiology and Immunology*, *374*, 189–209. https://doi.org/10.1007/82_2012_273.

Locke, J. C. W., & Elowitz, M. B. (2009). Using movies to analyse gene circuit dynamics in single cells. *Nature Reviews Microbiology*, *7*(5), 383–392. https://doi.org/10.1038/nrmicro2056.

Stinear, T. P., Seemann, T., Harrison, P. F., Jenkin, G. A., Davies, J. K., Johnson, P. D. R., et al. (2008). Insights from the complete genome sequence of Mycobacterium marinum on the evolution of Mycobacterium tuberculosis. *Genome Research*, *18*(5), 729–741. https://doi.org/10.1101/gr.075069.107.

Wakamoto, Y., Dhar, N., Chait, R., Schneider, K., Signorino-Gelo, F., Leibler, S., et al. (2013). Dynamic persistence of antibiotic-stressed mycobacteria. *Science (New York, N.Y.)*, *339*(6115), 91–95. https://doi.org/10.1126/science.1229858.

Wakamoto, Y., Inoue, I., Moriguchi, H., & Yasuda, K. (2001). Analysis of single-cell differences by use of an on-chip microculture system and optical trapping. *Fresenius' Journal of Analytical Chemistry*, *371*(2), 276–281.

Mechano-chemostats to study the effects of compressive stress on yeast

12

L.J. Holt*, O. Hallatschek[†], M. Delarue[‡,1]

*Institute for Systems Genetics, New York University Langone Health, New York, NY, United States
[†]Department of Physics and Integrative Biology, University of California, Berkeley, CA, United States
[‡]MILE, Laboratory for Analysis and Architecture of Systems, CNRS, Toulouse, France
[1]Corresponding author: e-mail address: mdelarue@laas.fr

CHAPTER OUTLINE

Methods in Cell Biology, Volume 147, ISSN 0091-679X, https://doi.org/10.1016/bs.mcb.2018.06.010

Abstract

Cells need to act upon the elastic extracellular matrix and against steric constraints when pro-liferating in a confined environment, leading to the build-up, at the population level, of a com-pressive, growth-induced, mechanical stress. Compressive mechanical stresses are ubiquitous to any cell population growing in a spatially-constrained environment, such as microbes or most solid tumors. They remain understudied, in particular in microbial populations, due to the lack of tools available to researchers. Here, we present various mechano-chemostats: microfluidic de-vices developed to study microbes under pressure. A mechano-chemostat permits researchers to control the intensity of growth-induced pressure through the control of cell confinement, while keeping cells in a defined chemical environment. These versatile devices enable the interroga-tion of physiological parameters influenced by mechanical compression at the single cell level and set a standard for the study of growth-induced compressive stress.

1 INTRODUCTION

Cells habitually proliferate in a confined environment. For example, microbes often inhabit micrometer-sized pores (Warscheid & Braams, 2000), and mammalian cells such as solid tumors, may be confined inside biological tissues (Stylianopoulos et al., 2012). In this context, cells develop compressive, growth-induced, mechanical stress (Fig. 1). Despite their paramount importance (Alessandri et al., 2013; Delarue et al., 2013, 2014; Fernández-Sánchez et al., 2015; Helmlinger, Netti, Lichtenbeld, Melder, & Jain, 1997; Stylianopoulos et al., 2012; Tse et al., 2012), compressive mechanical stresses have been much less explored than other types of mechanical stress, for example tensile stresses (Butcher, Alliston, & Weaver, 2009; Engler, Sen, Sweeney, & Discher, 2006; Fletcher & Mullins, 2010; Huang, Chen, & Ingber, 1998; Levental et al., 2009; Northey, Przybyla, & Weaver, 2017; Paszek et al., 2005), owing in part to the technical challenges of confining cells, and in par-ticular microbes.

Several approaches were used in the past to study cells under confinement. In a pioneering study, Helmlinger and colleagues embedded mammalian cells in an agarose gel in order to study the impact of growth-induced pressure

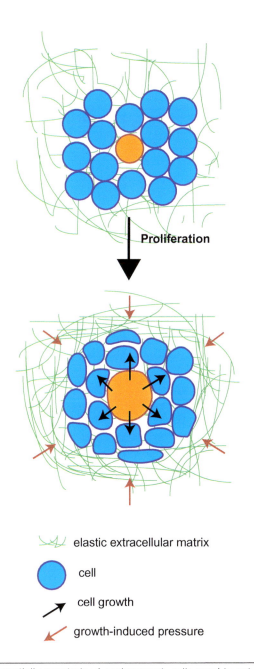

Proliferation

elastic extracellular matrix

cell

cell growth

growth-induced pressure

FIG. 1

When growing in a spatially-constrained environment, cells need to act upon their surroundings, for example other cells and an elastic matrix, to accommodate space for new cell material. Growth-induced contact pressures emerge naturally under these conditions.

(Helmlinger et al., 1997). This approach had drawbacks, in particular the poor characterization of the mechanical properties of the agarose gel, making the evaluation of growth-induced pressure complicated to assess. More recently, Alessandri and colleagues overcame these drawbacks by embedding mammalian cells in a fully-mechanically characterized alginate shell (Alessandri et al., 2013). However, both approaches cannot readily be transferred to the study of microbes under confinement because the dimensions and elasticity of hydrogels are not compatible with the biophysical characteristics of microbes. Notably the amount of growth-induced stress microbes can build up can be orders of magnitude higher than that of mammalian cells, in the 100 kPa range (Delarue et al., 2016) as opposed to the 1 kPa range (Dolega et al., 2017; Farge, 2003) for mammalian cells. Moreover, the cell population was fully confined in both methodologies, a situation that may not always resemble the natural habitat of microbes.

We have employed the large repertoire of tools offered by microfluidics (Autebert et al., 2012; Glawdel, Elbuken, Lee, & Ren, 2009; Groisman et al., 2005) to design several devices for the study of fungal cells under spatial confinement. Our device is complementary to previous methodologies that either deformed single *Saccharomyces cerevisiae* cells (Mishra et al., 2017), or partially confined *Schizosaccharomyces pombe* cells (Minc, Boudaoud, & Chang, 2009): they enable the study of cellular responses at the single cell level *in situ*. The model organism used to design these devices was the budding yeast *S. cerevisiae*, an organism of roughly 5 μm diameter with a thick elastic cell wall and a large turgor pressure of several bars (Hohmann, 2002). Our devices enable either active or passive modulation of spatial confinement, and can precisely control compressive stress, while keeping the chemical environment fixed. We termed these devices mechano-chemostats. This chapter is organized as follows: Section 2 presents the design and application of mechano-chemostats; in Section 3, we discuss the differences between passive and active devices, as well as how to measure growth-induced pressure; in Section 4, some results obtained with these devices are presented; we finally discuss in Section 5 potential future directions to create mechano-chemostats for other species.

2 PRESENTATION AND MICROFABRICATION OF THE MECHANO-CHEMOSTAT
2.1 PRESENTATION OF THE DEVICE

The mechano-chemostat consists of a confining chamber into which cells are loaded and can proliferate normally. The confining chamber is connected to a set of narrow channels too small for cells to enter (Fig. 2): These channels are used for media influx and efflux, with a media replacement time of tens of seconds allowing for constant control of the chemical environment while maintaining cell confinement. Several rounds of cell division quickly fill the whole chamber which, at a size of

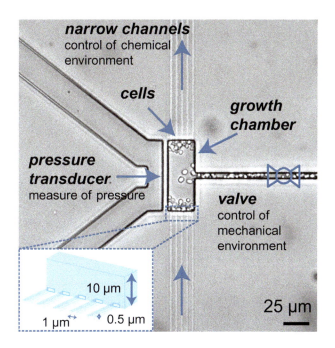

FIG. 2

Design of a mechano-chemostat. Cells are cultured in a confining chamber that is connected to a set of narrow channels that are too small for cells to enter. These channels are used to set the chemical environment. Cell outflow is limited by a valve that partially or totally confines the population, thus modulating growth-induced compression, which is measured by the pressure transducer. The valves can be active or passive, as illustrated in Fig. 5.

$65 \, \mu m \times 25 \, \mu m \times 10 \, \mu m$ (length × width × height), can accommodate about 100 cells. When the chamber is fully filled with cells, the excess cells flow out of the chamber through a microfluidic valve. We designed different sets of valves, passive or active, that are detailed in Section 3. These microfluidic "faucets" tune cell confinement in the confining chamber: when open, there is no confinement thus no growth-induced pressure, whereas when fully closed, total confinement results in the build-up of strong growth-induced compressive stress. Pressure is either measured directly or inferred from the deformation of the confining chamber.

2.2 FABRICATION PROTOCOL

The microfabrication protocol is classic photolithography (Campo & Greiner, 2007; Jo, Van Lerberghe, Motsegood, & Beebe, 2000) summarized below (Fig. 3). The material listed is specific to the fabrication of the devices presented in this chapter. We do not list the particular brand for the mask aligner or for the spin coater.

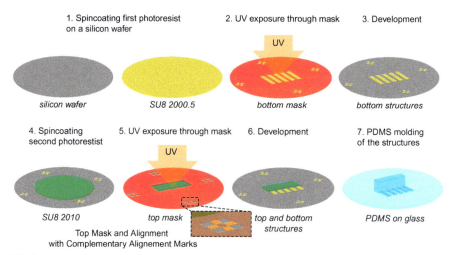

FIG. 3

Microfabrication steps. Classical photolithography is used to create silicon molds in two steps: the thin layer that defines geometry including nutrient channels first, then the second thicker layer that completes the growth chambers and input channels.

2.2.1 Material and equipment

- Negative photoresist, SU8 2000.5 and SU8 2010 (Microchem)
- SU8 developer (Microchem)
- Isopropanol
- Acetone
- Polydimethylsiloxane and curing agent (Sylgard 184)
- Number 1 thickness glass
- Puncher 0.75 mm (Harris)
- Reactive Ion Etcher (RIE, Diener)
- Oven set at 60 °C

2.2.2 Fabrication process

1. A first layer of SU8 2000.5 negative photoresist is spin-coated onto a silicon wafer to attain 0.5 μm of thickness, then pre-baked. Note that the baking temperatures and times depend on the thickness and on the type of photoresist used.
2. This first layer is insulated by UV illumination in a mask aligner, followed by a post-exposure bake and development. Attaining the 1 μm resolution of the nutrient channels is challenging and that the parameters usually used with the SU8 photoresists will have to be adjusted. In particular, we find that a smaller exposure and vacuum contact are often needed.
3. The first layer is developed using SU8 developer and then washed with isopropanol.

4. A second layer of SU8 2010 is spin-coated on the wafer to attain a 10 μm height. The edges are removed manually with a clean room paper and some acetone, followed by pre-baking. Again, pre-baking depends on the photoresist used.
5. Insulation of the second layer, followed by post-exposure bake.
6. The second layer is developed using and the top structures and alignment are checked under a microscope. If the structures are good, hard baking is follows. The wafer is silanized with trichlorosilane by vapor deposition to prevent attachment from PDMS.
7. 1:10 Sylgard 184 PDMS is finally molded on the silicon wafer, holes are created with the puncher, and the PDMS is bound to a #1 thickness glass by oxygen plasma activation with a reactive ion etcher (RIE) reactor. Typical parameters used are: 100 W, $P_{O_2} = 260$ mbar, exposure time of 30 s. When binding the PDMS to the glass, do not press it against it, as this may result in structures collapsing. Quickly place the device in an oven at 60 °C for at least 20 min.

Note that the optimization of step 7 is crucial and may depend on the RIE used. Particularly important parameters to adjust are the partial pressure of oxygen and the reaction time. You can ensure that the binding is optimal by trying to rip it manually. When the binding is good, it is not possible to detach PDMS from the glass and you should end up breaking the PDMS and observing traces of it on the glass slide.

2.3 LOADING CELLS AND IMAGING IN THE DEVICE

We present in the following how to load the device and the type of imaging that can be performed.

2.3.1 Material and equipment
- PDMS device prepared in Section 2.2
- Cells grown overnight in exponential phase
- SCD medium
- Sterile 1 mL syringes (Terumo)
- Sterile 23G blunt needles (Strategic Applications Inc.)
- PTFE microbore tubing 0.22 × 0.42 in. (Cole-Parmer)

2.3.2 Loading of the device
After connecting one of the main inlets to a syringe containing an overnight culture of cells at an OD ≈ 0.4, connect one of the two inlets for nutrients to a syringe containing the desired culture medium, and the other one to a tube to dispose of the waste (Fig. 4A). The loading of the device has to be performed in a specific sequence to avoid introduction of any air bubbles (Fig. 4B):

1. Apply pressure on the chemical syringe to fill both the outlet and the cell inlet with medium.
2. While keeping some pressure on the chemical syringe to avoid air coming back into this channel, apply pressure on the cell syringe, to chase the air through the outlet. At some point, air will be flowing out of the device through the outlet.

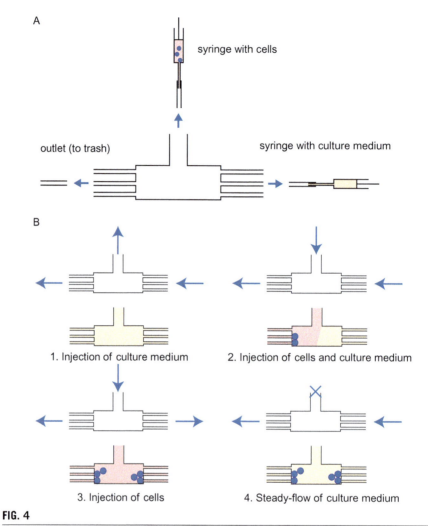

A

syringe with cells

outlet (to trash)

syringe with culture medium

B

1. Injection of culture medium

2. Injection of cells and culture medium

3. Injection of cells

4. Steady-flow of culture medium

FIG. 4

Loading of the device has to occur in a correct order. (A) Plug the syringes in the correct holes. (B) Load the medium first, then the cells, to fully get rid of air.

3. Only apply pressure on the cell syringe, to load the device with the desired amount of cells. Most of our experiments started with one to five cells. Loading the device can be stressful for the cells, so starting the experiment with just a few cells and allowing them to recover while they are filling the chamber ensures better reproducibility.

4. When the device is loaded, take out the cell syringe and its tubing, and seal this inlet with a plug, for instance made out of a melted tubing. Apply a steady flow that can be imposed with your favorite syringe pump, through the chemical inlet: because the cell inlet is closed, the medium will mainly flow through the chamber and the outlet.

2.3.3 Imaging

All of our devices are bound to #1 thickness glass. PDMS being transparent, this allows for imaging, both in bright field and in fluorescence. We have used magnification from $40\times$ to $100\times$, and imaging methods such as bright field, phase contrast, epifluorescence, confocal microscopy, fluorescence correlation spectroscopy (FCS), fluorescence recovery after photobleaching (FRAP), and even performed laser-ablation *in situ*. Because we can control the chemical environment, most of the classical biological staining techniques, such as immunofluorescence staining or FISH are also compatible with our devices and allow for precise analysis of the response to compressive stress. Hence, there is no particular method associated with imaging.

3 CONTROL OF THE MECHANICAL ENVIRONMENT

We designed valves of different kinds to limit the cell outflow from the confining chamber to impose a given degree of confinement to the cell population. Passive valves limit the flow without the use for external drive, in contrast with active valves which are water-driven pneumatics.

3.1 PASSIVE VALVES

3.1.1 Material and equipment

No particular equipment is needed to use the passive valves.

3.1.2 Using the passive valves

We designed a set of various obstacles that naturally limit cell outflow (Fig. 5A). We found that obstacles work by inducing jamming in the cell population (Delarue et al., 2016): force chains spanning the whole population stabilize it, leading to partial confinement and a build-up of growth-induced pressure. Pressure rises until the force network is broken, leading to an avalanche of cells through the outlet and a decrease of pressure. Over longer time periods, the geometry of the obstacle sets the mean steady-state pressure around which the cell population will stochastically vary. We also took advantage of the propensity of *S. cerevisiae* cells to jam to design a valve that fully confines the cell population (Fig. 5B). When cells fill the chamber, growth-induced pressure builds up through jamming and results in the deformation of PDMS. By adding side-pockets that surround the outlet channel to the confining chamber, PDMS deformation will result in these side-pockets pinching the exit channel, completely confining cells. This total confinement results in stronger jamming, and a larger pressure build-up, up to a stall pressure for cell growth.

Hence, the use of these valves to impose a growth-induced compressive stress does not require any external drive, and can be used with the devices prepared in Section 2 without further equipment. The deformation of the device informs on the amount of compressive stress (see Section 3.3).

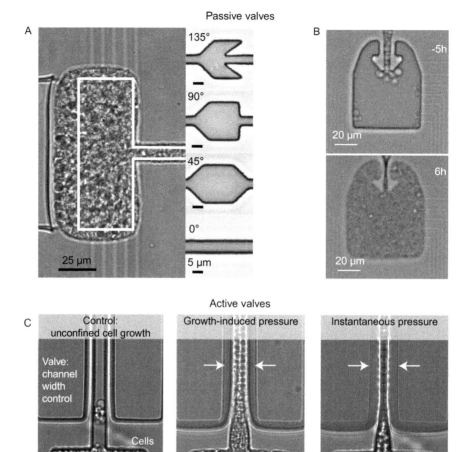

FIG. 5

Different types of valves are connected to the confining chamber. (A) Passive valves with various geometries will limit the flow and increase pressure through self-driven jamming. (B) A self-closing valve can be used to fully confine the cell population. (C) Vertical membranes can be actuated by a water-driven pressure to close the outlet channel to a desired amount, resulting in the build-up of growth-induced pressure. The pressure transducer can be used to either measure pressure through its deformation or used as a micropiston to further compress the cell population.

3.2 ACTIVE VALVES
3.2.1 Material and equipment
- Loaded device (see Section 2)
- Syringe pump (Cetoni GmbH)
- Pressure sensor (Cetoni GmbH)

3.2.2 Using the active valves
The approach with the side-pockets surrounding the outlet can be extended to design an externally water-driven valve. We surrounded the outlet with two vertical membranes that can be actuated with a measured water-driven pressure (Fig. 5C). The range of pressure being up to several bars, we chose to use a syringe pump coupled to a pressure sensor to impose a given pressure knowing the volume flown in the chamber, instead of a pressure pump that is limited in the maximum amount of pressure that can be applied when compressing air.

The deformation of the membranes by water-driven pressure can be used to pinch the outlet channel. When the outlet channel is open, the cell population is not confined and no pressure builds up. Conversely, when the channel is closed, cell outflow becomes limited and growth accumulates in the chamber, resulting in a progressive growth-induced pressure. We found a linear correlation between the closing of the valve and the intensity of growth-induced pressure (Delarue et al., 2017). This valve thus functions as a microfluidic "faucet." This type of valve is useful to precisely and dynamically control the intensity of growth-induced pressure in the chamber.

3.3 MEASURING PRESSURE
3.3.1 Material and equipment
- Loaded device (see Section 2)
- Syringe pump (Cetoni GmbH)
- Pressure sensor (Cetoni GmbH)

3.3.2 Pressure measurement
The confining chamber is in contact with a vertical membrane, a "pressure transducer" (Fig. 2) deformed by the build-up of growth-induced pressure. Growth-induced pressure can be measured in two ways:

1. Careful calibration of the deformation of the membrane by a given water-driven pressure before the experiment enables calculation of the pressure developed by the cell population by measuring the deformation of the membrane. We find a linear relationship between the deformation δ and the pressure P for a given Young's modulus E: $\delta = 12.09\, P/E$ in μm (Delarue et al., 2016).

2. The water-driven pressure applied on the membrane can be adjusted to keep the membrane at the same position at all time: when at mechanical equilibrium, the water-driven pressure matches the growth-induced pressure, providing a direct readout of the mechanical stress developed by the cell population.

Both modes present advantages and drawbacks. The first mode requires measurement of the mechanical properties of the PDMS beforehand but allows for rapid measurement of pressure changes that may occur through sudden avalanches of the cell population. In contrast, the second mode does not require any calibration, but requires a complex feedback loop where software has to control both the camera for image acquisition and the pressure pumps. We used Matlab to perform these operations, which is compatible with most commercial cameras and pressure pump drivers.

3.4 MICROPISTON AND INSTANTANEOUS COMPRESSIVE STRESS

3.4.1 Material and equipment
- Loaded device (see Section 2)
- Syringe pump (Cetoni GmbH)
- Pressure sensor (Cetoni GmbH)

3.4.2 Applying an instantaneous pressure
We used the pressure transducer to directly and instantaneously compress the cell population (Delarue et al., 2017): By increasing the water-driven pressure, one can deform the membrane in order to compress the cell population (Fig. 5C). Note that here the pressure that the cells experience does not directly correspond the pressure applied to the pressure transducer, as the pressure experienced would depend on the elastic properties of the PDMS and the packing fraction of cells. Hence pressure sensors such as the ones described in Dolega et al. (2017) should be used in combination to this method in order to accurately measure the intensity of this instantaneous pressure.

3.5 PASSIVE OR ACTIVE VALVES?

Each type of device has pros and cons. The passive valve devices are fast and easy-to-use and enable a large range of growth-induced pressure through the use of outlets of various geometries. We have successfully parallelized passive valve devices to enable high-throughput analysis (Delarue et al., 2017). Their disadvantage over the active valves is the lack of dynamic control of the pressure: once the cells in a device are enclosed by a passive valve, pressure can only build up to an extent imposed by the geometry of the valve. The use of active valves requires more specialized equipment like pressure or syringe pumps and hardware control. However, their versatility is excellent: growth-induced pressure can be dynamically controlled, where different pressure intensities can be tested during the same experiment on the same cell population to test for mechanical adaptation, and pressure can even be relaxed in order to study the reversibility of observed phenotypes.

4 PHYSIOLOGICAL RESPONSE TO A MECHANICAL STRESS

In this section, we present two different aspects of the impact of mechanical stress on cell physiology that we obtained using the previously described mechano-chemostats.

4.1 GROWTH-INDUCED PRESSURE LIMITS BOTH CELL GROWTH AND CELL PROLIFERATION

Growth rate in the mechano-chemostat can be measured at a steady-state pressure by the flow rate of cells leaving the chamber. We measured that this rate decreases with increasing compressive stress (Fig. 6A). In parallel, we found that cells under pressure are delayed at the beginning of the cell cycle, in a cell cycle phase termed G1 (Fig. 6B). It is worthwhile noting that both behaviors have also been observed in compressed mammalian cells (Montel et al., 2011, 2012). Whereas the decrease in growth rate comes from either active biological regulation or passive physical response to compressive stress, the decrease in cell proliferation is a biological response, suggesting that these unicellular organisms are capable of sensing and responding to mechanical stress.

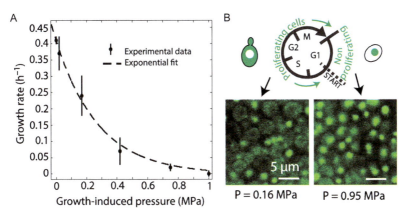

FIG. 6

Impact of compressive stresses on cell proliferation. (A) Growth rate decreases roughly exponentially as a function of growth-induced pressure. (B) Growth-induced pressure blocks the cells in the G1 phase of the cell cycle.

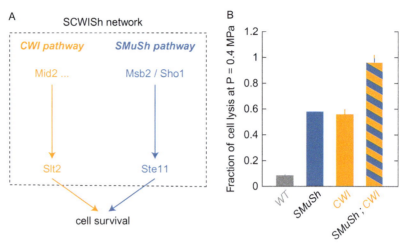

FIG. 7

The *SCWISh* network is essential for cell survival under pressure. (A) The mechano-sensitive *CWI* and *SMuSh* pathways define the *SCWISh* network. (B) Abrogation of both *CWI* and *SMuSh* pathways results in total cell death under compressive stress.

4.2 THE SCWISh NETWORK IS ESSENTIAL FOR SURVIVAL UNDER GROWTH-INDUCED PRESSURE

We used a high-throughput version of the passive devices in order to rapidly screen for mutants essential for cell survival under compressive stress. We found that the transmembrane mucin Msb2p, previously described as an osmosensor (Saito & Posas, 2012), was also a mechano-sensor of growth-induced compressive stress. Together with the scaffold protein Sho1p this mucin triggers the activation of the MAPKKK Ste11p to drive partial adaptation to compressive stress. We termed this pathway *SMuSh*, for Ste11p activation through mucin Msb2p and Sho1p (Fig. 7A). Both (Delarue et al., 2017; Mishra et al., 2017) additionally elucidated a role for the cell wall integrity pathway in the response to compressive stress. The cell wall integrity pathway (*CWI*) and *SMuSh* pathway act in parallel: abrogation of both *CWI* and *SMuSh* pathways totally abolishes cell survival under compressive stress (Fig. 7B). We named this network *SCWISh* (Survival through *CWI* and *SMuSh*). The *SCWISh* network is essential for cells to arrest in G1 and survive under mechanical pressure.

5 DISCUSSION

The proposed mechano-chemostats are extremely flexible microfluidic devices that enable the study of compressive mechanical stress. They are compatible with optical imaging, and the control of the chemical environment makes it possible to perform staining like FISH or immunofluorescence *in situ*. Moreover, high-throughput devices can be used to rapidly screen for multiple conditions, either genetic or chemical, that alter cell proliferation/survival under a compressive mechanical stress.

Even though our mechano-chemostats were developed for the fungus *S. cerevisiae*, we were also able to use them to study other yeasts including *S. pombe* and *Candida albicans*. The geometry of mechano-chemostats, notably the valve and the size of the nutrient channels, can either be downscaled to study smaller microbial organisms, or upscaled for the study of mammalian cells. The advantage of mechano-chemostats over existing techniques (Alessandri et al., 2013; Helmlinger et al., 1997; Minc et al., 2009; Mishra et al., 2017) resides in the versatility of their usage. They allow dynamic control of pressure, along with real-time control of the chemical environment. We believe that our approach could be adapted to ask a variety of fundamental questions regarding the impact of compressive stress on living organisms.

REFERENCES

Alessandri, K., Sarangi, B. R., Gurchenkov, V. V., Sinha, B., Kießling, T. R., Fetler, L., et al. (2013). Cellular capsules as a tool for multicellular spheroid production and for investigating the mechanics of tumor progression in vitro. *Proceedings of the National Academy of Sciences of the United States of America*, *110*(37), 14843–14848. https://doi.org/10.1073/pnas.1309482110.

Autebert, J., Coudert, B., Bidard, F.-C., Pierga, J.-Y., Descroix, S., Malaquin, L., et al. (2012). Microfluidic: An innovative tool for efficient cell sorting. *Methods*, *57*(3), 297–307.

Butcher, D. T., Alliston, T., & Weaver, V. M. (2009). A tense situation: Forcing tumour progression. *Nature Reviews Cancer*, *9*(2), 108–122.

Campo, A. d., & Greiner, C. (2007). SU-8: A photoresist for high-aspect-ratio and 3D submicron lithography. *Journal of Micromechanics and Microengineering*, *17*(6), R81–R95. https://doi.org/10.1088/0960-1317/17/6/R01.

Delarue, M., Hartung, J., Schreck, C., Gniewek, P., Hu, L., Herminghaus, S., et al. (2016). Self-driven jamming in growing microbial populations. *Nature Physics*, *12*(8), 762–766. https://doi.org/10.1038/nphys3741.

Delarue, M., Montel, F., Caen, O., Elgeti, J., Siaugue, J.-M., Vignjevic, D., et al. (2013). Mechanical control of cell flow in multicellular spheroids. *Physical Review Letters*, *110*(13), 138103. https://doi.org/10.1103/PhysRevLett.110.138103.

Delarue, M., Montel, F., Vignjevic, D., Prost, J., Joanny, J.-F., & Cappello, G. (2014). Compressive stress inhibits proliferation in tumor spheroids through a volume limitation. *Biophysical Journal*, *107*(8), 1821–1828. https://doi.org/10.1016/j.bpj.2014.08.031.

Delarue, M., Poterewicz, G., Hoxha, O., Choi, J., Yoo, W., Kayser, J., et al. (2017). SCWISh network is essential for survival under mechanical pressure. *Proceedings of the National Academy of Sciences of the United States of America*, *114*(51), 13465–13470. https://doi.org/10.1073/pnas.1711204114.

Dolega, M. E., Delarue, M., Ingremeau, F., Prost, J., Delon, A., & Cappello, G. (2017). Cell-like pressure sensors reveal increase of mechanical stress towards the core of multicellular spheroids under compression. *Nature Communications*, *8*. 14056. https://doi.org/10.1038/ncomms14056.

Engler, A. J., Sen, S., Sweeney, H. L., & Discher, D. E. (2006). Matrix elasticity directs stem cell lineage specification. *Cell*, *126*(4), 677–689.

Farge, E. (2003). Mechanical induction of Twist in the Drosophila foregut/stomodeal primordium. *Current Biology*, *13*(16), 1365–1377.

Fernández-Sánchez, M. E., Barbier, S., Whitehead, J., Béalle, G., Michel, A., Latorre-Ossa, H., et al. (2015). Mechanical induction of the tumorigenic β-catenin pathway by tumour growth pressure. *Nature*, *523*(7558), 92–95.

Fletcher, D. A., & Mullins, R. D. (2010). Cell mechanics and the cytoskeleton. *Nature*, *463*(7280), 485–492. https://doi.org/10.1038/nature08908.

Glawdel, T., Elbuken, C., Lee, L. E. J., & Ren, C. L. (2009). Microfluidic system with integrated electroosmotic pumps, concentration gradient generator and fish cell line (RTgill-W1)—Towards water toxicity testing. *Lab on a Chip*, *9*(22), 3243–3250. https://doi.org/10.1039/b911412m.

Groisman, A., Lobo, C., Cho, H., Campbell, J. K., Dufour, Y. S., Stevens, A. M., et al. (2005). A microfluidic chemostat for experiments with bacterial and yeast cells. *Nature Methods*, *2*(9), 685–689. https://doi.org/10.1038/nmeth784.

Helmlinger, G., Netti, P. A., Lichtenbeld, H. C., Melder, R. J., & Jain, R. K. (1997). Solid stress inhibits the growth of multicellular tumor spheroids. *Nature Biotechnology*, *15*, 778–783.

Hohmann, S. (2002). Osmotic stress signaling and osmoadaptation in yeasts osmotic stress signaling and osmoadaptation in yeasts. *Microbiology and Molecular Biology Reviews*, *66*(2), 300–372. https://doi.org/10.1128/MMBR.66.2.300.

Huang, S., Chen, C. S., & Ingber, D. E. (1998). Control of cyclin D1, p27Kip1, and cell cycle progression in human capillary endothelial cells by cell shape and cytoskeletal tension. *Molecular Biology of the Cell*, *9*(11), 3179–3193. https://doi.org/10.1091/mbc.9.11.3179.

Jo, B.-H., Van Lerberghe, L. M., Motsegood, K. M., & Beebe, D. J. (2000). Three-dimensional micro-channel fabrication in polydimethylsiloxane (PDMS) elastomer. *Journal of Microelectromechanical Systems*, *9*(1), 76–81. https://doi.org/10.1109/84.825780.

Levental, K. R., Yu, H., Kass, L., Lakins, J. N., Egeblad, M., Erler, J. T., et al. (2009). Matrix crosslinking forces tumor progression by enhancing integrin signaling. *Cell*, *139*(5), 891–906. https://doi.org/10.1016/j.cell.2009.10.027.

Minc, N., Boudaoud, A., & Chang, F. (2009). Mechanical forces of fission yeast growth. *Current Biology*, *19*(13), 1096–1101. https://doi.org/10.1016/j.cub.2009.05.031.

Mishra, R., van Drogen, F., Dechant, R., Oh, S., Jeon, N. L., Lee, S. S., et al. (2017). Protein kinase C and calcineurin cooperatively mediate cell survival under compressive mechanical stress. *Proceedings of the National Academy of Sciences of the United States of America*, *114*(51), 13471–13476. https://doi.org/10.1073/pnas.1709079114.

Montel, F., Delarue, M., Elgeti, J., Malaquin, L., Basan, M., Risler, T., et al. (2011). Stress clamp experiments on multicellular tumor spheroids. *Physical Review Letters*, *107*(18), 188102. https://doi.org/10.1103/PhysRevLett.107.188102.

Montel, F., Delarue, M., Elgeti, J., Vignjevic, D., Cappello, G., & Prost, J. (2012). Isotropic stress reduces cell proliferation in tumor spheroids. *New Journal of Physics*, *14*(5), 55008. https://doi.org/10.1088/1367-2630/14/5/055008.

Northey, J. J., Przybyla, L., & Weaver, V. M. (2017). Tissue force programs cell fate and tumor aggression. *Cancer Discovery*, *7*(11), 1224–1237. https://doi.org/10.1158/2159-8290.CD-16-0733.

Paszek, M. J., Zahir, N., Johnson, K. R., Lakins, J. N., Rozenberg, G. I., Gefen, A., et al. (2005). Tensional homeostasis and the malignant phenotype. *Cancer Cell*, *8*(3), 241–254. https://doi.org/10.1016/j.ccr.2005.08.010.

Saito, H., & Posas, F. (2012). Response to hyperosmotic stress. *Genetics*, *192*(2), 289–318. https://doi.org/10.1534/genetics.112.140863.

Stylianopoulos, T., Martin, J. D., Chauhan, V. P., Jain, S. R., Diop-Frimpong, B., Bardeesy, N., et al. (2012). Causes, consequences, and remedies for growth-induced solid stress in murine and human tumors. *Proceedings of the National Academy of Sciences of the United States of America*, *109*(38), 15101–15108. https://doi.org/10.1073/pnas.1213353109.

Tse, J. M., Cheng, G., Tyrrell, J. A., Wilcox-Adelman, S. A., Boucher, Y., Jain, R. K., et al. (2012). Mechanical compression drives cancer cells toward invasive phenotype. *Proceedings of the National Academy of Sciences of the United States of America*, *109*(3), 911–916. https://doi.org/10.1073/pnas.1118910109.

Warscheid, T., & Braams, J. (2000). Biodeterioration of stone: A review. *International Biodeterioration & Biodegradation*, *46*(4), 343–368. https://doi.org/10.1016/S0964-8305(00)00109-8.